Techniques from Vedic Math to improve mental memory, speed and accuracy of calculations

First Edition

Pankaj Sharma, M Tech (CSE),PMP,PRINCE2

Neetu Sharma, MSC (Computer Science)

For suggestions and Feedback, Contact No-919810996356

ISBN-13: 978-1475086430 (CreateSpace-Assigned)
ISBN-10: 1475086431

PREFACE

This Book presents techniques taken from ancient Vedic system that helps to do calculations much faster than the normal techniques. The book is helpful for students in school as well as for those planning to appear in competitive exams such as SAT, GMAT, CAT, NTSE and Olympiad. To demonstrate the practical application sample problems from these exams are solved using Vedic techniques in this book. It's also useful for senior executives from the corporate sector who have to deal with numbers a lot and strength in this area is considered as an important attribute for them.

Vedic Mathematics is based on Atharvaveda which was rediscovered from the Vedas between 1911 and 1918 by Sri Bharati Krsna Tirthaji (1884-1960). Vedic Mathematics is based on sixteen sutras which are called as formulas or techniques that are easy to learn and apply. These principles are general in nature and can be applied in many ways. In practice many applications of the sutras may be learned and combined to solve actual problems. Research shows that application of Vedic mathematics uses both the part of the brain and this helps in improving the mental memory.

Albeit the techniques from Vedic mathematics aids quick calculations in the all the scenarios, however, the techniques are highly relevant for pattern based calculations and increases the speed by manifold. In this book we have put a lot of emphasis on Vedic techniques for Addition subtraction, multiplication, division, divisibility, square/square root, cube/cube root and fractions

About Author

Pankaj Sharma is a post graduate in Computer Engineering from Birla Institute of Technology and Science Pilani and is a certified PMP (PMI –USA) and Prince 2 (APMG-UK) professional. He is extremely passionate about Vedic Mathematics, has been doing research and conducting workshops on Vedic mathematics for more than a decade. He has practically seen the benefits of learning Vedic mathematics on students and professionals in Asian countries. Through this book he would like to support broader audience.

Neetu Sharma is a post graduate in Computer Science has been conducting workshops for students aspiring for Olympiad, NTSE and GMAT exam for more than a decade. She is highly enthusiastic about disseminating knowledge on Vedic Mathematics and spends most of her time in doing research on the subject. Before starting her career as a trainer she has worked with corporate such as Bharati Group and Headstrong Consulting

Contents

Insight

This section Illustrates some of the techniques from Vedic Mathematics that makes calculations at least 10 times faster than the normal methods.

Multiplication of numbers near the base

Vedic techniques for multiplying the number close to the base are atleast 10 times faster than the normal method. The techniques can be used to multiply large numbers such as 999999 x 999998, 10003 x 10004. We have shown the examples along with the steps below

Multiply 88 by 98.

88 12

98 02

86 24

Both 88 and 98 are close to 100. 88 is 12 below 100 and 98 is 2 below 100. (100- 88 = 12 & 100-98 = 02)

As shown above **86** comes from subtracting crosswise: 88 - 2 = 86 (or 98 - 12 = 86): you can subtract either way, you will always get the same answer. And the **24** in the answer is just 12 x 2: you multiply vertically. Since our base is 100, so the figure at the right hand side should consist of only two digits.

So **88 x 98 = 8624**

$$103 \quad +3$$
$$107 \quad +7$$
$$11021$$

Let see how this works in case of bigger numbers **(9999 X 9997)** through examples below

9999 0001

9996 0003 .

9997 0003

9996 0003

The steps are as follows

1. Write the two numbers to be multiplied above and below.

2. Take a base for the calculation. The base should be that power of 10, which is nearest to the number to be multiplied.

3. Subtract each number from the base and write the remainder with respective plus (+) or minus (-) sign.

4. The product will have two parts left and right.

5. The left hand product is obtained by cross operation of the numbers written diagonally.

6. The right hand product is obtained by multiplying the two digits written at right hand portion

Multiplying a number with 11, 111 or more no of ones

Multiplication by 11

To multiply any 2-figure number by 11 we just put the total of the two figures between the 2 figures.

- **26 x 11 = 286**

Notice that the outer figures in 286 are the 26 being multiplied. And the middle figure is just 2 and 6 added up.

- **So 72 x 11 = 792**

For bigger numbers

234 x 11 = 2574

We put the 2 and the 4 at the ends.
We add the first pair 2 + 3 = 5.
And we add the last pair: 3 + 4 = 7.

To multiply a two-digit number by 111, add the two digits and if the sum is a single digit, write this digit Two Times in between the original digits of the number. Some examples:

- **36 x 111= 3996**
- **54 x 111= 5994**

The same idea works if the sum of the two digits is not a single digit, but you should write down the last digit of the sum twice, but remember to carry if needed as illusrated in example below

$65 \times 111 = 7215$

- **57 x 111= <u>6327</u>** $87 \times 111 \ 9657$

Because 5+7=12, but then you have to carry the one twice.

For 3 digit numbers
Carry if any of these sums is more than one digit.
Thus 123x111 = 1 | 3 (=1+2) | 6 (=1+2+3) | 5 (=2+3) | 3

Similarly,

- **241 x 111= <u>26751</u>** 26751

For an example where carrying is needed

- **352 x 111= <u>39072</u>**

3 | 8 (=3+5) | 10 (=3+5+2)| 7 (=5+2)| 2 39072
= 3 | 8 | 10 | 7 | 2 = 3 | 9 | 0 | 7 | 2
= 39072

481×111
$= 53391$ 53391

677×111
$= 75147$

391×111
$= 43401$

Multiplying numbers where the first figures are the same and the last figures add up to 10.

- **32 x 38 = <u>1216</u>**

    (3 x (3+1)) | (2 x 8) = (3 x 4) | (2 x 8) = 1216

Both numbers here start with 3 and the last figures (2 and 8) add up to 10.

So we just multiply 3 by 4 (the next number up) to get **12** for the first part of the answer. And we multiply the last figures: 2 x 8 = **16** to get the last part of the answer.

*| is just a separator. Left hand side denotes tens place, right hand side denotes units place

Diagrammatically:

$$3 \; 2 \; \times \; 3 \; 8 \; = \; 1 \; 2 \quad 1 \; 6$$

(annotations: $2 \times 8 = 16$; $3 \times 4 = 12$)

- **81 x 89 = <u>7209</u>**

    8 x 9 | 1 x 9 = 7209

    We put 09 since we need two figures as in all the other examples.

Cube Root of a number with less than 7 digits

Example1: Cube root 35937

1. Place a bar over the digits of the number from right to left, leaving 2 digits at a time

$$\overline{3} \quad \overline{5} \ 9 \overline{3 \ 7}$$

2. The unit digit with bar is 7. Therefore the unit digit of the cube root is 3 (as 3^3 =27)

3. The next bar falls on 5. The tens digit of the number is largest number whose cube root is less than or equal to 35 and as $3^3 < 3 < 4^3$

4. Hence the cube root is 33

Multiplying with decimal numbers of the format 1.1, 2.2, 3.3 etc

523 X 1.1 = 523 X (1+.1) = 523 + 52.3 = 575.3

523 X 2.2 = 523 X (2 + .2) = 4106 + 410.6= 4516.6

Similarly you can work out the calculations for 3.3, 4.4, and 5.5 and so on.

This is extremely useful in many real time scenarios. For Example if you have to convert Kilograms into pounds.

Convert 52 Kilogram into Pounds

As 1 Kilogram = 2.2 Pounds

52 Kilograms = 52 X 2.2 Pounds = 104 + 10.4 = 124.4

Chapter-1

Vedic math improves mental memory and helps in improving the speed of calculation by making you smart in dealing with numbers. Before learning the Vedic techniques for dealing with complex calculations, let's understand the basics of the number system.

Basics of number system

Number system includes Decimal, Binary and Hexadecimal System. Since the objective is to learn smart tricks for quick calculations , In this book we will focus only on the Decimal number system; the other systems are beyond the scope of the book.

The decimal system has a base of 10; every digit is multiplied by 10 raised to the power corresponding to that digit's position. Few are some of the examples.

The unit digit is multiplied by 1, tens by 10 and 100th by 100 and so on so forth.

$93 = 9 \times 10 + 3$;

$933 = 9 \times 100 + 3 \times 10 + 3$;

Types available in Decimal number system

Prime numbers

Prime numbers are only divisible by one and themselves. Like 1, 2, 3, 5, 7, 11, 13, 17….

Co-prime numbers

Pair of numbers that have no common factors other than 1 are Co- Prime numbers.

Example: 15 and 23, 8 and 21.

Composite Numbers

Composite numbers are the number that can be split into factors. Example 12 can be split into factors,

3 X 4.

Odd numbers

An integer which is not multiple of two is odd number, example 21, 27.

Complements of a Number

When we subtract a number from its nearest base of ten which is more than the number, we get the complement of that number. Example, with the no 86, the nearest base of ten more than 86 is 100.

100 – 86 =14 is the complement of 86. This is very useful for performing calculations when the numbers are close to the base. Example below illustrates that:

98 + 99 + 97 = 100 – 2 + 100 – 1 + 100 -3 = 300 – 6 = 294

For larger numbers a quick way to find the complement is by subtracting all the digits from nine and last from ten.

Complement of 576,

Take each of the digits from 9 and the last from ten

5 from 9 = 4

7 from 9 = 2

6 from 10= 4, the complement is 424

---

6 7 8

3 2 2

Complement of 70408,

Take each of the digits from 9 and the last from ten

7 from 9 = 2

0 from 9 = 9

4 from 9 = 5

0 from 9 = 9

8 from 10= 2

Exercise

Write down the complements of the following

87   65   123987   8004   3487   99995   6666221   5063
478902   8637

13   35   876013   6513   3333779

1996   00005   4037

521098

Digital Sum

Digit sum also called as Digital root of a number is found by adding the digits in a number and adding again if necessary until a single figure is reached. Below are some examples,

Digit sum of 72 is 7 + 2 = 9.

Similarly for 21302 we get 2+1+3+0+2 = 8.

For 89 we first get 8+9 = 17, but 1+7 = 8, so digit sum of 89 is 8.

The key benefit of digital sum is to quickly validate your calculations. This is based on the rules explained below

Digit sum or digital root is useful for divisibility testing (Divisibility by 3 and 9) and for checking calculations. Following are some of the key characteristics of Digital sums

1. A number is divisible by 3 and 9 if their digital sums are divisible by 3 and 9

2. Digital sum of the result obtained after Addition, Subtraction, Multiplication and Division of numbers is equal to the digital sum of the number obtained if we perform the similar operation with their digital sum.

The digital root of a number is also the remainder we get when that number is divided by nine.

Addition o f 232 + 332 = 564, the digital sum of 564 is 6

Digital Sum of 232 + Digital Sum of 332 = 7 + 8 =15 =6, this value is same as the digital sum of result obtained after adding the actual numbers

232 + 332 – 100 = 464, the digital sum is 5

Digital Sum of 232 + Digital Sum of 332 – Digital Sum of 100 = 7 + 8 – 1=5, this value is same as the digital sum of result obtained after adding the actual numbers

102 X108 = 11016, the digital sum of the product is 9

Digital sum of (102) X Digital Sum of (108) = 3 X 9 = 9, this value is same as the result obtained after multiplying the actual numbers.

Dividend = 876543, divisor = 123, quotient = 7126 and remainder = 45

The digital roots are 6 for 8765432, 6 for 123, 7 for 7126 and 9 for 45.

Since Divisor x Quotient + Remainder = 6 x 7 +9 =42 =6 which should be equal to the digital root of the dividend, provides us with a check to correctness

20

9 Point Circle

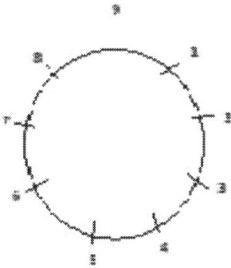

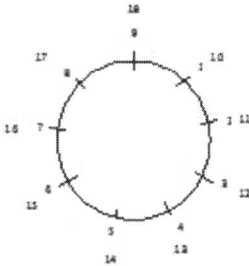

The behavior of Digital sums can be described through the 9 point circle shown above, when we move clockwise starting from 1 after reaching 9 the digital sum of the next number starts from 1. Example after 9 the number is 10 and its Digital sum is 1, after 18 the number is 19 and its digital sum is 1. This means that while finding the digital sum we can simply ignore the digit with value 9 or any digits adding up to 9. As illustrated in example below

Digit sum of 5999 is just 5, we simply delete 9.

If there is nothing left after casting out 9s then digital root is 9.

Exercise (Validate the following using digital sum)

1) 7534 + 2459 + 1932 + 6547 = 16472 2) 112065 + 360085 + 289872 + 1563345 = 918367 3) 3746735 − 2837546 = 909189

4) 588 x 512 = 301056   5) 966 x 973 = 939918

6) 13579 ÷ 975, q = 13, r = 904   7) 11199171 ÷ 99979, q = 112, r = 1523

8) 876542 − 32538 − 698547 = 145447

1234
37
10
① + 614
+   1 1   +
    ②   17 = 11 + 11
        + ⑧   = ④   12 + 2 + 1
                        5 5 6
7 1 1 2   = ⑪   ⑥
  8 3

Addition of large numbers

Techniques in Vedic Mathematics help in quick addition of large numbers by relieving you from the pain of dealing with big numbers and the carry over . The stepwise procedure for adding numbers is stated below.

1. Add the unit digits column wise downwards.
2. When the running total becomes greater than 10 put the dot to the number next to the digit in the preceding column.
3. Move ahead with the excess of 10 and add it to the next digit of the column.
4. The presence of a dot on a number will make it one richer by itself.
5. When the sum of the digits is more than 10 and there is no column left then put a zero preceding the number so that dot can be placed over it.

Example: 87643 + 84397 + 38549 + 29765 =

	5th	4th	3rd	2nd	1st
	8	7	6	4	3
	·	·	·	·	·
0	8	4	3	9	7
	·	·			
0	3	8	5	4	9
		·	·	·	·
+	2	9	7	6	5
2	4	0	3	5	4

Explanation

In first column 3 + 7 = 10, a dot is put on 9 in the preceding column and the excess 0 (10-10= 0) is added to 9 and 9+5 =14, a dot is put on 6 in the preceding column and the excess 4 (14-10= 4) is taken down.

In second column 4 + ˙9 (=10) = 14, so a dot is put on 3 in the preceding column and the excess 4 (14-10= 4) is added to 4, 4

+ 4= 8 and 8+˙6 (=7) = 15 and the excess 5 (15-10= 5) is taken down.

In third column 6+˙3 (=4) = 10, a dot is put on 4 in the preceding column and the excess 0 (10-10= 0) is added to5, 5 + ˙7 (=8) =13, a dot is put on 9 in the preceding column and the excess 3(13-10= 3) is taken down.

In fourth column 7 + ˙4 (5)= 12, a dot is put on 8 in the preceding column and the excess 2 (12-10= 2) is added to 8, 2 + 8=10, a dot is put on 3 in the preceding column and the excess 0 (10-10= 0) is added to ˙9(=10), 0 + 10=10, a dot is put on 2 in the preceding column and 0 is taken down.

In fifth column 8 + ˙8 (=9) = 17, so a dot needs to be put to the preceding number in sixth column hence zero is placed in sixth column and excess 7 (17-10= 7) is added to ˙3(=4), 7 + 4= 11, a dot needs to be put to the preceding number in sixth column hence zero is placed in sixth column again and the excess 1 (11-10= 1) is added to ˙2(=3), 1 + 3 =4, 4 is taken down.

Number Splitting (<u>make pair of 2-digit numbers from right</u>)

This is the very useful device for splitting a difficult sum into two or more easy ones.

For quick mental sums number splitting can considerably reduce the work involved in a calculation.

Procedure

1. Take at least two-digit numbers in each intersection of the rows and columns.
2. While summing up the individual column, bring one number as the multiple of 10.
3. The excess should be added to the next number of that column.
4. When sum is greater than 100, place a dot on the digit next to it in the preceding column as done in above method.
5. Follow the same procedure for the remaining column.

Example: 2345 + 6738 =

Now split the sum into two parts, each part can be done easily and mentally;

$$2^{nd} \quad 1^{st}$$

$$2\,3 \mid 4\,5$$

$$\underline{6\,7 \mid 3\,8} +$$

$$\underline{9\,0 \mid 8\,3}$$

Explanation

In first column: 45 + 38 = 40 + 5 + 30 + 8

$$= (40 + 30) + 5 + 8$$

$$= 70 + 13 = 83, \text{ write 83 in the column below the line}$$

In second column: 23 + 67 = 20 + 3 + 60 + 7

$$= (20 + 60) + 3 + 7$$

$$= 80 + 10 = 90, \text{ write 90 in the column below the}$$
line

Example: 254 + 368 + 596=

	2nd	1st

$$
\begin{array}{r r r}
 & 2^{\text{nd}} & 1^{\text{st}} \\
 & 2 & 54 \\
 & \cdot & \\
 & 3 & 68 \\
 & \cdot & \\
+ & 5 \quad | & 96 \\
\hline
 & 12 & 18
\end{array}
$$

Explanation

In the first column: $54 + 68 = 50 + 4 + 60 + 8$

$$= (50+60) + 4 + 8$$

$$= 110 + 12 = 122$$

Since $122 > 100$, so place a dot over 3 and add the excess 22(122-100= 22) to the next number 96.

$$22 + 96 = 20 + 2 + 90 + 6$$

$$= (20+90) + 2 + 6$$

$$= 110 + 8 = 118$$

Since 118 > 100, so place a dot over 5 and write the excess 18(118-100) in the column below the line.

In the second column: 2 + ˙3 + ˙5= 2 + 4 + 6= 12 will be placed in the column below the line.

The above discussed Vedic method for addition can be used for other type of sums as well.

- Kilogram- grams
- Rupee- paisa
- Decimal
- Hour- minute

Example: Add 112 Kg 65 g, 360 Kg 85 g, 284 Kg 678 g, 154 Kg 469 g and 72 Kg 424 g

Make two columns; Kg and g. Since 1 Kg = 1000g, so place three-digit number under gram column.

Kg	g
112	065
	..
360	085
. .	.
284	678
.. .	.
154	469
	..
+ 72	424
983	721

Explanation:

Write 65g and 85g as 065 and 085. Start adding from the unit column. 5 + 5=10; so place a dot over 8

0 + 8 + 9 = 17>10, place a dot over 6 and 7 + 4 = 11>10, place a dot over 2.

Write 1 to the unit column and proceed in the same way to find the sum of all the column.

Example: Add Rs 28 40 p, Rs 32 04p, Rs 65 78p, Rs 38 92p and Rs 16 92p

Rs	P
28	40
32	04
.	.
65	78
	.
38	92
	.
+ 16	92
182	06

Column 1:

40 + 04 + 78 = 40 + 4 +70 + 8

= (40 + 70) + 4 + 8 = 110 + 12 =112>100, the excess (122-100 = 22) and add it with the next number of the 1st column. Place dot over 5 in preceding column.

22 + 92 = 20 + 90 + 2 + 2

= 110 +4 = 114>100, the excess (114-100 = 142) and add it with the next number of the 1st column. Place dot over 8 in preceding column.

14 + 92 = 10 + 4 + 90 + 2

= 100 + 6=106>100, so place dot over 6 in preceding column and write the excess as the answer of column 1

Column 2

28 + 32 + 65˙ = 20 + 8 + 30 + 2 + 60 + 6

= 110 + 16 =126>100, place a dot over 0 as no digit left in second column and excess(126-100=26) will be added to next number of second column.

26 + 38˙ + 16˙= 20 + 6 + 30 + 9 + 10 + 7

= 60 + 22 = 82, take it down and 0˙=1, so answer of second column is 182

Exercise

Add the following
75934 + 87628 + 34879 + 14093 + 256
762874 + 593487 + 4789 + 187049
3476928 + 274039 + 8752684 + 48998
Rs 659   38p + Rs 968   15p + Rs 159 64p + Rs6 98p + Rs 98   36p

VINCULUMS or BAR NUMBERS

Vinculum of a number is obtained by subtracting the nearest working base (which is greater than the number) and its difference from the nearest working base. Example Vinculum of 232 is 240 (Nearest working base) – 8 (difference from the nearest working base); this is 240-8 and is represented as 248 (Bar over eight). Vinculums help in simplifying the calculations when we have to deal with the numbers with bigger unit digits. Suppose we have to add 29+39 it is easier to use vinculums of these numbers 30-1 and 40-1 to find the sum instead of original numbers as shown below.

$$29 + 39 = 30-1+40-1= 68$$

Above was a simple case in the subsequent section of the book you will find out how vinculums are useful in complex calculations.

From the above examples, we can conclude the following

1. To change the digit into a bar number, that digit is replaced by its complement and the digit to the left is increased by one.

2. To change a vinculum digit back into its ordinary form, write the complement of the bar number and subtract 1 from next digit to the left.

Exercise

Convert each units, tens digit into bar numbers.

59   78   89   458   109   9989   7739   8888   328   7716

Exercise

Convert back to their ordinary form

1) 5$\bar{1}$ **2)** 8$\bar{4}$ **3)** 7$\bar{8}$ **4)** 71$\bar{2}$ **5)** 61$\bar{1}$ **6)** 85$\bar{2}$ **7)** 46$\bar{5}$

8) 39$\bar{4}$

If a number has a bar number in it then split the number after the bar

Example: 5$\bar{1}$3, split the number into two parts: 5$\bar{1}$ / 3

   The complement of 1 is 9 and 5-1 = 4 (3 remains unchanged).

   So 5$\bar{1}$ = 49, the answer is 493.

Similarly 6$\bar{2}$ 1, split the number into two parts: 6$\bar{2}$ / 1

   6$\bar{2}$ = 58, the answer is 581

$54\bar{2}8$, split the number as $5\,/\,4\bar{2}\,/\,8$

$4\bar{2} = 38$, the answer is 5388

Numbers may have more than one vinculum number.

Example: $3\bar{2}45\bar{1}2$, split the number as $3\bar{2}\,/\,4\,/\,5\bar{1}\,/\,2$

$3\bar{2} = 28$ and $5\bar{1} = 49$, so the answer is 284492

$3\bar{1}3\bar{2}3\bar{3}$, split the number as $3\bar{1}\,/\,3\bar{2}\,/\,3\bar{3}$

$3\bar{1} = 29$, $3\bar{2} = 28$ and $3\bar{3} = 27$, so the answer is 292827

Exercise

Convert each bar number into their ordinary form
$4\bar{3}3 \quad 12\bar{1} \quad 67\bar{2}8 \quad 2\bar{2}2\bar{2}22 \quad 31\bar{4}53 \quad 6\bar{2}7\bar{3} \quad 7\bar{1}52$
$\bar{4}1\bar{1}10 \quad\quad 6\bar{5}3\bar{2}1 \quad\quad 544\bar{1}6$

Number may have two or more vinculum numbers together.

Example: $5\overline{33}$, the complement of 33 is 67 and 5 -1 = 4.

So the answer is 467.

$26\overline{21}$, the complement of 21 is 79 and 6 -1 = 5

So the answer is 2579.

$7\overline{02}$, the complement of 02 is 98 and 7 -1 = 6

So the answer is 698.

$4\overline{20}$, the complement of 20 is 80 and 4 -1 = 3

So the answer is 380.

Exercise

Convert the bar numbers
)$6\overline{12}$ $7\overline{33}$ $9\overline{04}$ $6\overline{12}4$ $5\overline{3}1 22$ $33\overline{22}44$ $6\overline{123}$ $3\overline{3}14$ $\overline{1}$ $4\overline{2}1\overline{22}$ $6\overline{3322}$ $5\overline{1}046\overline{1}7$ $4\overline{1}29\overline{704}$

SUBTRACTION

Vedic method discussed here will help you to ease the very concept of subtraction and even the difficult looking subtraction can be done in a few seconds mentally.

Use the formula ALL FROM 9 AND THE LAST FROM 10 to perform instant subtractions

- For example **1000 - 357 = 643**

    We simply take each figure in 357 from 9 and the last figure from 10.

    ```
    1  0  0  0   -   3   5   7
                     |   |   |
                 from 9 from 9 from 10
                     ↓   ↓   ↓
                 =   6   4   3
    ```
    So the answer is **1000 - 357 = 643**

    This always works for subtractions from numbers consisting of a 1 followed by noughts: 100; 1000; 10,000 etc.

- Similarly **10,000 - 1049 = 8951**
    ```
    1  0 , 0  0  0   -   1   0   4   9
                         |   |   |   |
                     from 9 from 9 from 9 from 10
                         ↓   ↓   ↓   ↓
                     =   8   9   5   1
    ```

- For 1000 - 83, in which we have more zeros than figures in the numbers being subtracted, we simply suppose 83 is 083.

    So **1000 - 83** becomes **1000 - 083** = <u>**917**</u>

Exercise

1) 1000 – 777   **2)** 1000 – 283   **3)** 1000 – 505   **4)** 10,000 – 2345

5) 10000 – 9876   **6)** 100 – 57   **7)** 1000 – 57   **8)** 10,000 – 321

9) 10,000 – 38   **10)** 10,000 – 1101

One less than previous one and Complementary method (ALL FROM 9 AND THE LAST FROM 10)

Right to left

In simple case when all of the top row digits are greater or equal to the digits below then each digit is subtracted from the one above. Complements are used when this not the case. The basic method is to take the difference of the two digits and when the bottom row digit is larger, write down the complement of the difference. When complements are no longer needed we subtract an extra 1 from the next left-hand column.

Example

Subtract **3876** from **5322**.

Starting from right, 6 is more than 2, so we take the difference i.e. 4 and write its complement from 10 (since it is the last), that is 6	5 3 2 2 3 8 7 6
In the next column, the difference between 7 and 2 is 5 and the complement (from 9) is 4.	6 5 3 2 2 3 8 7 6
In the next column, the difference between 8 and 3 is 5 and the complement (from 9) is 4.	4 6
In the next column, 5 is greater than and so we can finish using complements. This is done by reducing the answer by 1 after ordinary subtraction that is 5 – 3 – 1 = 1	5 3 2 2 3 8 7 6 4 4 6
So, the answer is 1446.	5 3 2 2 3 8 7 6 1 4 4 6

Alternatively we can use bar numbers to represent negative numbers and convert them into the normal number to get the result

Exercise

2569347 – 1598635

187956896476325 – 146210987315672

658997 – 467925

8976543 – 2498672

479867 – 364780

598023 – 489790

Chapter -2

We often come across complex multiplication (Many in real life scenarios) while dealing with problems related to unit & currency conversion , mensuration and commercial math. The techniques provided in this chapter are really helpful in simplifying and improving your ability to multiply . The techniques are largely based on pattern identification and are at least ten times faster than the conventional methods. We would like to emphasize that after learning these techniques try to practice them mentally without taking the aid of pen and paper. This will definitely help you in improving your mental memory. In the sections below we have explained the techniques to be used based on the patterns of the number to be multiplied.

Multiplication of numbers near a base

This is useful when the numbers to be multiplied are close to the base such as 99,98,91 ,999,997,99998,11,109,1009 etc. This is also handy when one of the number to be multiplied is close to the base such as 99 x 21

The steps for multiplying two numbers close to the base are as follows.

1. Write the two numbers to be multiplied above and below.

2. Take a base for the calculation. The base should be that power of 10, which is nearest to the number to be multiplied.

3. Subtract each number from the base and write the remainder with respective plus (+) or minus (-) sign.

4. The product will have two parts left and right.

5. The left hand product is obtained by cross operation of the numbers written diagonally.

6. The right hand product is obtained by multiplying the two digits written at right hand portion.

Example:

Multiplying numbers close to the base of 100 (Both the numbers less than the base)

- **Multiply 88 by 98.**

Both 88 and 98 are close to 100. 88 is 12 below 100 and 98 is 2 below 100. (100- 88 = 12 & 100-98 = 02)

$$
\begin{array}{r}
88 - 12 \\
\times\; 1 \\
98 -\;\; 2 \\
\hline
86\;\;\; 24 \\
\end{array}
$$

As before the **86** comes from subtracting crosswise: 88 - 2 = 86 (or 98 - 12 = 86): you can subtract either way, you will always get the same answer. And the **24** in the answer is just 12 x 2: you multiply vertically. Since our base is 100, so the figure at the right hand side should consist of only two digits.

So **88 x 98 = 8624**

- **Multiply 96 by 72.**

Both 96 and 72 are close to 100. 96 is 04 below 100 (04 as the base is 100) and 72 is 28 below 100. (100- 96 = 4 & 100-72 = 28)

(100)

96 - 04

× |

72 - 28

68 $_1$12 = 6912

As before the **68** comes from subtracting crosswise: 96 - 28 = 68 (or 72 - 4 = 68): you can subtract either way, you will always get the same answer. And the **112** in the answer is just 28 x 4: you multiply vertically.

Here there is a carry: the 1 in the 112 goes over to make 68 into 69. Since our base is 100, so the figure at the right hand side should consist of only two digits.

Exercise

Multiply these
1) 89 x 96   **2)** 88 x 94   **3)** 77 x 98   **4)** 93 x 96
5) 94 x 92   **6)** 94 x 99   **7)** 98 x 97

Numbers above the base of 100

- **103 x 104 = 10712**

The answer is in two parts: 107 and 12, 107 is just 103 + 4 (or 104 + 3),
and 12 is just 3 x 4.

- **107 x 106 = 11342**

   107 + 6 = 113 and 7 x 6 = 42

Multiply these
1) 102 x 107   **2)** 106 x 103   **3)** 104 x 104
4) 109 x 108   **5)** 101 x123   **6)** 103 x102

Multiplying numbers where one is above and the other is below the base 100.

44

- **108 x 94 = <u>10152</u>**

Now 108 is 8 above 100 and 94 is 6 below 100.

(100)	(100)	(100)
108 + 08	108 + 08	108 + 08
× |	× |	× |
<u>94 - 06</u>	<u>94 - 06</u>	<u>94 - 06</u>
<u>102 - 48</u>	<u>102 $\overline{48}$</u>	<u>101 52</u> = 10152 answ

As before the **102** comes from subtracting crosswise: 108 - 06 = 102 (or from adding crosswise 94 + 08 = 102): you can operate either way, you will always get the same answer. And the product of +8 and -6 is -48 and this is set down as $\overline{48}$.

To devinculate the $\overline{48}$ we take 1 away from the digit immediately to the left, that is 2 – 1= 1, and use All from nine and last from ten on 48 to obtain 52 (i.e., subtract 48 from base 100).

Devinculate means whenever the right hand product is with the negative sign subtract 1 from the left hand figure and subtract the right hand figure from the base

---

Multiply these

1) 102 x 89   **2)** 104 x 97   **3)** 104 x 98   **4)** 109 x 85

5) 101 x 88   **6)** 103 x 96   **7)** 106 x 99   **8)** 112 x 86

---

9) 115 x 94   **10)** 121 x 92

Multiplying numbers that are closer to the base of 1000

- **Multiply 888 by 997.**

(1000)

888 - 112

× |

997 - 003

885 336

- **Multiply 594 by 995** (carry over)

(1000)	(1000)	(1000)
594 - 406	594 - 406	594 - 406
×		×		×	
995 - 005	995 - 005	995 - 005
	589 $_2$030	591 030

46

Since the working base for above example is 1000, so the figure at the right hand side should not be more than three digits hence the 2 in the 2030 goes to the left to make 589 into 591.

Multiply these

1) 878x976   **2)** 954x690   **3)** 943x803   **4)** 845x999   **5)** 896x947   **6)** 9996x9998

7) 9978x8967   **8)** 9845x9767   **9)** 9990x9898   **10)** 8979x9542   **11)** 1008x1005

12) 1308x1007   **13)** 1500x1008   **14)** 1234x1007   **15)** 1005x1235   **16)** 1006x987

17) 1209x869   **18)** 1206x995   **19)**1004x879   **20)**1205x950

Proportionately

This method is applicable only when the numbers are very far from the base. We take two types of base: theoretical base (taken in power of 10) and working base (taken in multiple of 10). Here we take base which falls nearer to the number, according to our convenience.

- **48 x 42 = <u>2016</u>**

Theoretical base is 10 and working base is 10 x 4= 40

$$48 + 8 \qquad\qquad 48 + 8 \qquad\qquad 48 + 8$$

$$\underline{\text{x } 42 + 2} \qquad \underline{\text{x } \quad 42 + 2} \qquad \underline{\text{x } 42 + 2}$$

$$50 \quad 16 \qquad\qquad 4 \text{ x } 50 \quad {}_1 6 \qquad 200 \quad {}_1 6 = 2016$$

In order to get the final answer we have to multiply the left-hand side by the multiple of working base (4 in this case). The number of digits in right-hand part of the answer corresponds to the base 10; therefore 1 digit on the right-hand and extra digit is carried over to the left side after it is multiplied with multiple of working base.

- **71 x 87 = <u>6177</u>**

$$71 - 9 \qquad\qquad 71 - 9 \qquad\qquad 71 - 9 \qquad\qquad 71 - 9$$

$$\underline{\text{x } 87 + 7} \qquad \underline{\text{x } 87 + 7} \qquad \underline{\text{x } 87 + 7} \qquad \underline{\text{x } 87 + 7}$$

$$\underline{78} \ \ -63 \qquad 8 \text{ x } \underline{78} \ \ -63 \qquad \underline{624} \ \ {}_{-6}\text{-}3 \qquad \underline{618} \ \ \text{-}3$$

$$71 - 9$$

$$\underline{\text{x } 87 + 7}$$

$$617 \quad 7$$

48

Here there is a negative carry: the -6 in the -63 goes over to make 624 into 618. Since our base is 10, so the figure at the right hand side should consist of only one digit and -3 will become 7 (10-3) and 1 is reduced from left-hand side.

- **252 x 299 = 75348**

Theoretical base is 100 and working base is 100 x 3= 300

252 - 48	252 - 48	252 - 48
x 299 - 01	x 299 - 01	x 299 - 01

- **213 x 203 = 43239**

Theoretical base is 100 and working base is 100 x 2= 300

213 + 13	213 + 13	213 + 13
x 203 + 03	x 203 + 03	x 203 + 03

Multiply these

42x41   204x207   321x303   199x198   312x307   44x56
5003x5108

Multiplication by 11

To multiply any 2-figure number by 11 we just put the total of the two figures between the 2 figures.

- **26 x 11 = 286**

Notice that the outer figures in 286 are the 26 being multiplied. And the middle figure is just 2 and 6 added up.

- **So 72 x 11 = 792**

Multiply by 11:

1) 43   2) 81   3) 15   4) 44   5) 11

- **77 x 11 = 847** (carry over)

This involves a carry figure because 7 + 7 = 14, we get 77 x 11 = 7_147 = 847.

Multiply by 11:

1) 88   2) 84   3) 48   4) 73   5) 56

- **234 x 11 = 2574**

    We put the 2 and the 4 at the ends.
    We add the first pair 2 + 3 = 5.
    And we add the last pair: 3 + 4 = 7.

Multiply by 11:

1) 151   2) 527   3) 333   4) 714   5) 909

- **13423 x 11 = <u>147653</u>**

  We put the 1 and the 3 at the ends.
  We add the first pair: 1 + 3 = 4.
  Then add the next pair: 3 + 4 = 7.

  And then add the adjacent pair: 4 + 2 = 6.

  And then add the last pair: 2 + 3 = 5

Multiply by 11:

1) 125342   2) 249214   3) 523154   4) 614231   5) 825362

MULTIPLYING A NUMBER BY 111

To multiply a two-digit number by 111, add the two digits and if the sum is a single digit, write this digit Two Times in between the original digits of the number. Some examples:

- **36 x 111= <u>3996</u>**
- **54 x 111= <u>5994</u>**

The same idea works if the sum of the two digits is not a single digit, but you should write down the last digit of the sum twice, but remember to carry if needed as illusrated in example below

- **57 x 111= <u>6327</u>**

Because 5+7=12, but then you have to carry the one twice.

For 3 digit numbers
Carry if any of these sums is more than one digit.
Thus 123x111 = 1 | 3 (=1+2) | 6 (=1+2+3) | 5 (=2+3) | 3

Similarly,

- **241 x 111= <u>26751</u>**

For an example where carrying is needed

- **352 x 111= <u>39072</u>**

    3 | 8 (=3+5) | 10 (=3+5+2)| 7 (=5+2)| 2
    = 3 | 8 | 10 | 7 | 2 = 3 | 9 | 0 | 7 | 2
    = 39072

Multiply these

59x111   89x111   659x111   748x111   46x101   246x1001
439x1001   53x10101 616x101

The numbers of 9's- 99,999,9999,99999 ....)

Step 1: Subtract 1 from the number other than 9's and put it on the left side.

Step2: The digit at the right hand side is obtained on subtracting the remainder at the left from the number of 9's.

- **12x 99 = <u>1188</u>**

    1 2

    x 9 9

    12 99

    -1 -11

    <u>11 88</u>

- **4365x 9999 = <u>43645635</u>**

$$4\ 3\ 6\ 5$$

$$\underline{x\ 9\ 9\ 9\ 9}$$

4365 9999

$$\underline{-1\ -4364}$$

4364 5635

When digits of 9's numbers is more than the other number.

Step 1: Subtract 1 from the number other than 9's and put it on the left side.

Step2: The digit at the right hand side is obtained on subtracting the remainder at the left from the number of 9's.

- **44x 999 = <u>43956</u>**

$$4\ 4$$

$$\underline{x\ 9\ 9\ 9}$$

44 999

$$\underline{-1-\ 43}$$

43 956

54

- **3458 x 999999 = <u>347996542</u>**

$$3\ 4\ 5\ 8$$

$$\underline{x\ 9\ 9\ 9\ 9\ 9\ 9}$$

$$3458\ 999999$$

$$\underline{-1\ \ -\ 3457}$$

$$\underline{3457\ \ 996542}$$

When digits of 9's numbers is less than the other number

Step 1: Put as many zeros as the number of 9's to the other number.

Step2: Subtract the original number from the figure obtained in step 1.

- **1564 x 99 = <u>154836</u>**

1564	156400
x <u>9 9</u>	<u>- 1564</u>
	<u>154836</u>

1564 x 99 = 154836

- **783459 x 9999 = <u>7833806541</u>**

$$\begin{array}{r} 783459 \\ \times\,\underline{9999} \end{array}$$

$$\begin{array}{r} 7834590000 \\ -\quad\underline{783459} \\ 7833806541 \end{array}$$

Multiply these

1234x9999	456x99	65498x99999	3487x99	435x99999
89x999	397x999	67x9999	4823x999	

Multiplying numbers where the first figures are the same and the last figures add up to 10.

- **32 x 38 = <u>1216</u>**

    $(3 \times (3+1)) \mid (2 \times 8) = (3 \times 4) \mid (2 \times 8) = 1216$

Both numbers here start with 3 and the last figures (2 and 8) add up to 10.

So we just multiply 3 by 4 (the next number up) to get **12** for the first part of the answer. And we multiply the last figures: 2 x 8 = **16** to get the last part of the answer.

*| is just a separator. Left hand side denotes tens place, right hand side denotes units place

Diagrammatically:

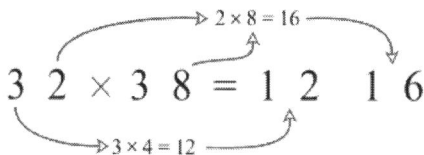

$$3\ 2 \times 3\ 8 = 1\ 2\ 1\ 6$$

with $2 \times 8 = 16$ and $3 \times 4 = 12$

- **81 x 89 = <u>7209</u>**

  8 x 9 | 1 x 9 = 7209

  We put 09 since we need two figures as in all the other examples.

- **44 x 46 = <u>2024</u>**

  = (4 x 5) | (4 x 6) = 2024

- **37 x 33 = <u>1221</u>**

  = (3 x 4) | (7 x 3) = 1221

- **11 x 19 = <u>209</u>**

  = (1 x 2) | (1 x 9) = 209

  Also it can be extended such as

- **292 x 208 = <u>60736</u>**

  Here 92 + 08 = 100, L.H.S portion is same i.e. 2

= (2 x 3) x 10 | 92 x 8 (Note: if 3 digit numbers are multiplied, L.H.S has to be multiplied by 10)

60 | 736 (for 100 raise the L.H.S. product by 0) = 60736.

- **848 x 852 = <u>722496</u>**

  Here 48 + 52 = 100,

  L.H.S portion is 8 and its next number is 9.

  = 8 x 9 x 10 | 48 x 52 (Note: For 48 x 52, use methods shown above)

  720 | 2496

  = 722496.

  [L.H.S product is to be multiplied by 10 and 2 to be carried over because the base is 100].

- **693 x 607**

  = 6 x 7 x 10 | 93 x 7 = 420 | 651 = 420651.

Multiply these

1) 43 x 47   **2)** 24 x 26   **3)** 62 x 68   **4)** 59 x 51
      5) 77 x 73

6) 742 x 758   **7)** 394 x 306

Multiplying numbers where the bases are different.

- **9988 x 77 = <u>7688276</u>**

Here the numbers are close to different bases: 10,000 and 100, and 9988 is 12 below 10,000 and 77 is 23 below 100.we imagine or write it as shown below. It is important to align the numbers as shown because the 23 is not subtracted from 88 but from the 99 above the 77. Then multiply the right hand side: 12 x 23 = 276. Note that the number of digits in right-hand part of the answer corresponds to the base of lower number (base 100, therefore 2 digits on the right and extra digit is carried over to the left side.

9988 – 12	9988 - 12	9988 - 12
<u>77 - 23</u>	<u>77 - 23</u>	<u>77 - 23</u>
7688 276	**7688 ₂76**	**7690 76** answer

- **9998 x 94 = <u>939812</u>**

9998 – 02

<u>94   - 06</u>

9398   12 answer

- **889 x 9998 = <u>8888222</u>**

  889 – 111

  <u>9998 - 002</u>

  <u>8888   222 </u>answer

- **10007 x 1003 = <u>10037021</u>**

  10007 + 007

  <u>1003   + 003</u>

  <u>10037   021</u>

Here 003 is added to 1000 above the 1003. Then multiply the right hand side: 7 x 3 = 21. Note that the number of digits in right-hand part of the answer corresponds to the base of lower number (base 1000, therefore 3 digits on the right side is required hence 0 is put in front of 21).

Multiply these

1) 97 x 998   **2)** 9988 x 98   **3)** 9996 x 988   **4)** 104 x 1017

5) 106 x 1015 **6)** 1122 x 105 **7)** 10034 x 102   **8)** 104 x 10105

Vertically and Crosswise Method

The diagram below will help in understanding and remembering the vertically and crosswise pattern. Each dot represents a digit in the number and the lines joining the dots stand for digits to be multiplied.

Multiplication of 2-digit numbers

Multiplication of 3-digit number

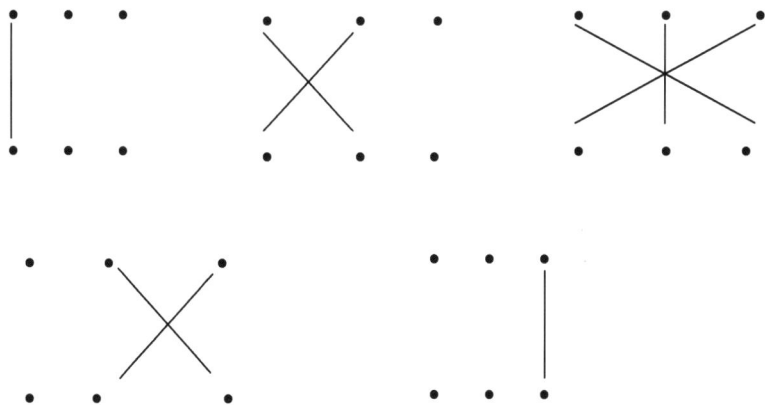

Multiplication of 4-digit number

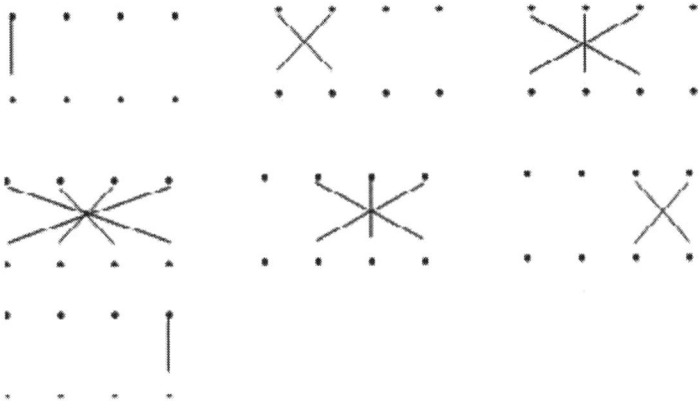

Multiply 2-digit numbers together

- **21 x 23 = <u>483</u>**

$$
\begin{array}{cc}
2 & 1 \\
| \times | \\
2 \quad\quad 3 \times \\
\hline
4 \quad 8 \quad 3
\end{array}
$$

There are 3 steps:

    A) Multiply **vertically on the right**: 1 x 3 = **3**.
       This gives the first figure of the answer.
    B) Multiply **crosswise and add**: 2 x 3 + 1 x 2 = **8**.
       This gives the middle figure.
    C) Multiply **vertically on the left**: 2 x 2 = **4**.
       This gives the last figure of the answer.

- Similarly **61 x 31 = <u>1891</u>**

$$
\begin{array}{ccc}
6 & & 1 \\
 & \times & \\
\underline{3} & & \underline{1} \times \\
18 & 9 & 1
\end{array}
$$

- 1 x 1 = **1**; 6 x 1 + 1 x 3 = **9**; 6 x 3 = **18**

Multiply these

1) 14 x 21   **2)** 22 x 31   **3)** 21 x 31   **4)** 21 x 22   **5)** 32 x 21

Multiply 2-digit numbers together (Involve carry over)

- **21 x 26 = <u>546</u>**

$$
\begin{array}{cc}
2 & 1 \\
| \times | \\
\hline
2 & 6 \times \\
\hline
4 \; {}_14 \; 6 & = 546
\end{array}
$$

A) Multiply **vertically on the right**: 1 x 6 = **6**.
   This gives the first figure of the answer.
B) Multiply **crosswise and add**: 2 x 6 + 1 x 2 = **14.**
   This gives the middle figure.
C) Multiply **vertically on the left**: 2 x 2 =**4.**
   This gives the last figure of the answer.

The method is the same as above except that we get a 2-figure number, 14, in the
middle step, so the 1 is carried over to the left (4 becomes 5). So the answer is 546

- **33 x 44 = <u>1452</u>**

There may be more than one carry

$$
\begin{array}{cc}
3 & 3 \\
| \times | \\
\hline
4 & 4 \times \\
\hline
12 \; {}_24 \; {}_12 & = 1452
\end{array}
$$

A) Multiply **vertically on the right**: 3 x 4= **12** and the 1 here is carried over to the middle.

B) Multiply **crosswise and add**: 3 x 4 + 3x 4 = **24+1**(carry over from the right side)= **25** and the 2 here is carried over to the left side

c) Multiply **vertically on the left**: 3 x 4 =**12 + 2**(carry over from the middle)= **14.**

Multiply these

1) 21 x 47   **2)** 23 x 43   **3)** 32 x 53   **4)** 42 x 32   **5)** 71 x 72   **6)** 32 x 56

7) 32 x 54   **8)** 31 x 72   **9)** 44 x 53   **10)** 54 x 64   **11)** 36 x 48   **12)** 35 x 59

Multiply 3-digit numbers together

- **362 x 134 = 48508**

Starting from right, the first answer digit is 2 x 4 = 8

```
  3 6 2
X 1 3 4
      8
```

The second step is the sum of the crosswise product of the four right-hand most digits, that is (6 x 4) + (2 x 3) = 24 + 6 = 30.0 is put down and there is 3 to carry.

```
  3 6 2
X 1 3 4
    0 8
```

The middle step is to add the crosswise product of all six digits in the following order, (3 x 4) + (6 x 3) + (2 x1) = 32, with the carry 3 this is 35. 5 is put down and there is 3 to carry.

3

362

X 1 3 4

5 0 8

The next step is the crosswise product of the four left-hand most digits, that is (3 x 3) + (6 x 1) = 15 and with the carry 3 this is 18. 8 is put down and there is 1 to carry.

3 3

362

X 1 3 4

8 5 0 8

The final step is the product of the two left-hand most digits, 3 x 1= 3 and with carry 1 this is 4

1 3 3

362

X 1 3 4

4 8 5 0 8

Multiply 4-digit numbers together

- **3142 x 5013 = <u>15750846</u>**

2 x 3 = 6

(4 x 3) +(2 x 1) = 14

(1 x 3) + (4 x 1) + (2 x 0) + 1(carry over)= 8

(3 x 3) + (1 x 1) + (4 x 0) + (2 x 5) = 20

(3 x 1) + (1 x 0) + (4 x 5) + 2(carry over) =25

(3 x 0) + (1 x 5) + 2(carry over) = 7

3 x 5 = 15

Multiply

412x312   144x162   423x203   789x121   512x370
909x131   324x78   45x433   1053x1041   2710x1321
5712x8320   9438x5741   7212x9021   5216x33
137x1280   11213x11321   14402x11232   23142x13
231204x112

Multiplication of 3 numbers

- **12 x 13 x 15 = <u>2340</u>**

Step 1: Write the difference of each number from its base. Place it against each number. For above example the base is 10 and differences from the base are 2, 3, 5.

Number	Difference from the base
12	+ 2
13	+ 3
15	+ 5

Step 2: Add all the differences to the base.

$$10 + 2 + 3 + 5 = 20$$

Step 3: start multiplying the differences in a pair of two at a time and add them.

$$(2 \times 3) + (2 \times 5) + (3 \times 5) = 31$$

Step 4: Multiply the differences.

$$2 \times 3 \times 5 = 30$$

Step 5: arrange the result obtained from above steps as shown below.

20   31   30 and adding the result from right to left in the direction of arrow, we get the answer.

$$2\,0 \quad 31 \quad 30$$

+ +

= 2340

- **105 x 104 x 109 = <u>1190280</u>**

Number	Difference from the base
105	+ 5
104	+ 4
109	+ 9

Base + differences= 100 + 5 + 4 + 9 = 118

Multiplying the differences in a pair of two add them = (5 x 4) + (4 x 9) + (5 x 9) = 101

Multiply differences = 5 x 4 x 9 = 180

Arrange the result obtained

$$1\,1\,8 \quad 1\,0\,1 \quad 1\,8\,0$$

+ +

= 1190280

- **989 x 995 x 1012 = <u>995863660</u>**

Here the base is 1000 so the differences from the base are:

Number	Difference from the base
989	-11

Multiply these

13x15x17   11x13x16   14x17x19   12x16x18   109x106x105
110x112x120   114x108x119   1009x1002x1007
1014x1016x1013   1008x986x945   1005x1013x978

995	- 5
1012	+ 12

Base + differences= 1000 -11 - 5 + 12 = 996

Multiplying the differences in a pair of two add them = (-11 x -5) + (-5 x +12) + (-11 x +12) = 55 – 60 – 132 = -137

Multiply differences = -11 x – 5 x 12 = 660

Arrange the result obtained from above steps

9 9 6   (-1 3 7) 6 6 0

9 9 5   (1000 – 137) 6 6 0

9 9 5 8 6 3 6 6 0

Chapter – 3

Divisibility is to check whether a number is completely divisible (leaves remainder as zero)) by other numbers or not. A good command on divisibility aids you in quick solutions to the problems in the areas such as mensuration, commercial math and also in dealing with calculations in real life scenarios.

Checking divisibility of numbers from numbers ranging from 2 to 11 and their multiples are quite simple and are explained in most of the text books. Checking divisibility with prime numbers such as 13,17,19, 29 and the bigger prime numbers is challenging. Vedic mathematics provide powerful techniques to determine the divisibility with these numbers using osculation. Osculation is described later in this chapter in the table below we have presented a quick snapshot of methods to determine divisibility from numbers ranging from 2 to 11.

Divisibility

Divisibility by 2	A number is divisible by 2 if it ends with an even number
Divisibility by 3	A number is divisible by three if it's digital sum is divisible by 3
Divisibility by 4	A number is divisible by four when the last two digits are divisible by four or when the sum of ultimate and twice the penultimate is divisible by four
Divisibility by 5	A number is divisible by 5 if it ends with five or nought

Divisibility by 6	A number is divisible by 6 when it is divisible by both 2 and 3
Divisibility by 7	Take the last digit in a number. Double and subtract the last digit in your number from the rest of the digits. Repeat the process for larger numbers. Example: 357 (Double the 7 to get 14. Subtract 14 from 35 to get 21 which is divisible by 7 and we can now say that 357 is divisible by 7
Divisibility by 8	A number is divisible by eight when the last three digits are divisible by eight or the sum of ultimate , twice the penultimate and four times the pen-penultimate is divisible by eight
Divisibility by 9	A number is divisible by nine when the digital root is nine
Divisibility by 10	A number is divisible by 10 if it ends with a nought
Divisibility by 11	A number is divisible by 11 if the sum of the digits in the odd places and the sum of the digits in the even places are equal or differ by a multiple of 11

Exercise

Check the divisibility of the following numbers with
2,3,4,5,6,7,8,9,10,11

256   3246   5368   2111034   6574602   128   9786   600
328886   646   3576   87231   57400

9350936   1286   2178   243558   3217000   5768574632
659418243   9272635443   10000123

8460153   17453   108   145   7299   9272635463   7685709
954   5433412   1587345

5151515151   72762   452452   8982982   0292736
242426   52252   5557775

Divisibility rules for composite numbers

To_obtain a divisibility rule for any composite number
such as 6, 12, 15 and 18 we use combination of the rule
for their factors.

Example:

A number is divisible by 6 when it is divisible by both 2 and 3

A number is divisible by 12 when it is divisible by both 3 and 4

A number is divisible by 15 when it is divisible by both 3 and 5

A number is divisible by 18 when it is divisible by both 2 and 9

A number is divisible by 24 when it is divisible by 3 and 8

A number is divisible by 30 when it is divisible by both 3 and 10

Exercise

Check the divisibility of the following numbers with 12, 15, 18 and 24

256   3246   5368   2111034   6574602   128   9786   600
328886   646   3576   87231

57400   9350936   1286   2178   243558   3217000
5768574632   659418243   9272635443

10000123   8460153   17453   108   145   7299
9272635463

Divisibility rules for larger prime numbers

We often face challenges in assessing the divisibility with number such as 13, 17, 19, 31 etc. Vedic math provides simple techniques for this using positive and negative osculation. Before starting with positive osculation lets understand what is Ekadhika

One more than the one before – Ekadhika

Ekadhika is the number "one more" than the one before when the number ends in a 9 or a series of 9's.

For example the Ekadhika for **19** is **2** because the one before the **9** is **1** and one more than **1** is **2**.

Similarly for **29** the Ekadhika is **3**.

For **13** the Ekadhika is **4** because **to obtain a 9 at the end we must multiply 13 by 3**, which gives 39 and for 39 Ekadhika is **4**

For **7** Ekadhika is **5** which is derived from **7 x 7 =49** and one before the **9** is **4** and one more than **4** is **5**.

For **27** Ekadhika is **19** which is derived from **27 x 7 =189** and one before the **9** is **18** and one more than **18** is **19**.

Exercise

Find the Ekadhika for the following
17   21   23   49   43   53   59   89   99

Osculation

There are two types of oscillators – Positive Osculator and Negative Osculator.

The Positive Osculator

Positive osculator is useful for testing the divisibility by a number which ends with 9. Positive Osculator is just the Ekadhika.

We osculate a number by multiplying its last figure by the osculator and adding the result to the previous figure.

Example

- Find whether 91 is divisible by 7.

The Ekadhika for 7 is 5, so we osculate the 91 with 5.

Multiply the 1 in 91 by the osculator, 5 and add the result to the 9.

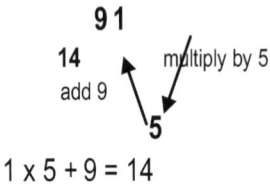

9 1

14   multiply by 5

add 9

5

$1 \times 5 + 9 = 14$

We get 14 as the result and 14 is clearly divisible by 7. Hence we can say that 91 is also divisible by 7.

- Test 78 for divisibility by 13.

Ekadhika for 13 (13 x 3=39) is 4 so osculate 78 by 4.

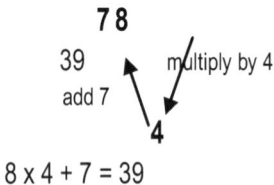

7 8

39   multiply by 4

add 7

4

$8 \times 4 + 7 = 39$

We get 39 as the result and 39 is clearly divisible by 13. Hence we can say that 78 is also divisible by 13.

- Test 86 divisible by 19

Ekadhika for 19 is 2, so osculate 86 by 2.

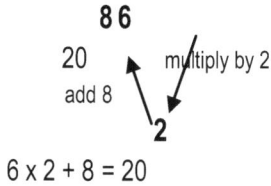

86

20   multiply by 2

add 8

2

$6 \times 2 + 8 = 20$

We get 20 as the result and 20 is not divisible by 19. Hence we can say that 86 is not divisible by 19.

Exercise

Test the divisibility of the following	
a) Test for the divisibility by 13:	**d) Test for the divisibility by 19:**
91 86 65 98 84 57 102	95 76 114 102 51 57 91
b) Test for the divisibility by 7:	**e) Test for the divisibility by 23:**
63 84 98 114 76 52 91	57 87 116 114 86 98 76
c) Test for the divisibility by 17:	
85 116 95 51 102 86 52	

Testing Longer Numbers

The osculation procedure is extended for testing the divisibility of a long number.

Example

- Test whether 247 is divisible by 19

The Ekadhika is 2. We multiply 7 by 2 and add it to the next figure, 4: So, 7 x 2 + 4 =18. We put this under 4 as shown

```
2   4   7
    18
```

```
2   4   7
19  18
```

Here 1 in the 18 is carried over to join 2. Now we multiply 8 by 2 and add 1 next to it and 2. So, 8 x 2 + 1 + 2 = 19

We end up with 19 and therefore 247 is divisible by 19

- Test whether 4617 is divisible by 19

The Ekadhika is 2. We multiply 7 by 2 and add it to 1:
  So, 7 x 2 + 1 =15.

```
4   6   1   7
        15
```

Then multiply 5 by 2 and add 1 and 6:

```
4   6   1   7
   17  15
```

Then multiply 7 by 2 and add 1 and 4:

```
4   6   1   7
19 17 15
```

We end up with 19 and therefore 4617 is divisible by 19

- Test whether 13455 is divisible by 23

The Ekadhika is 7(23 x 3 = 69). We osculate as above using 7:

1	3	4	5
5			
69	59	8	40

We got 69 at the end which is clearly divisible by 23. Therefore 13455 is divisible by 23

Exercise

Test the following numbers for divisibility by 19

779   4503   323   4313   589   2774   1996   10203   30201   14003   234555

Test the divisibility of following numbers

4321 by 109   10404 by 17   4173 by 13   5432 by 7   1003 by 59   4173 by 13

4802 by 49   2254 by 23   41963 by 29

Other Divisors

Example

- Is 6308 divisible by 38

Here, 38 = 2 x 19 so we have to test for the divisibility by 2 and 19. This number is clearly divisible by 2 so we need to test for 19. The osculator is 2:

$$6 \quad 3 \quad 0 \quad 8$$
$$19 \quad 16 \quad 16$$

6308 is also divisible by 19 so it is divisible by 38.

- Is 5572 divisible by 21

Here 21 = 3 x 7 so we have to test for 3 and 7. It is not divisible by 3 (as the digit sum is 1) so we don't proceed further. Hence 5572 is not divisible by 21.

- Is 1764 divisible by 28

Here 28 = 4 x 7, the last two figures of 1764 indicate that it is divisible by 4 so we test for 7 next.
The osculator is 5:

$$1 \quad 7 \quad 6 \quad 4$$
$$49 \quad 39 \quad 26$$

And we see that the test for 7 is passed. So 1764 is divisible by 28

- In testing for divisibility by a number we look at the factors of that number first and start with the easiest factors.

Exercise

Test the divisibility of the following
334455 by 39   21645 by 65   1771 by 46   3538 by 58   767 by 95   1254 by 38   305448 by 52   37932 by 58   5985 by 95

The Negative Osculator

Negative osculator is useful for testing the divisibility by a number which ends with 1. Negative osculator for these numbers is the figure which we get after dropping the 1 from that number.
For example the negative osculator for 31 is 3, we just drop the 1. Similarly, negative osculator for 41 is 4 and to get the negative osculator for 17 we need to get a 1 at the end of the number and this is done by multiplying 17 by 3 which is 51 and so the negative osculator for 17 is 5.

Exercise

Find the negative osculator
81   11   27   7   13   91   19   23   37   101 61

Example

- Is 3813 divisible by 31

The negative osculator here is 3. The osculation process is little bit different for negative osculator.

We begin by putting a bar over every other figure in 3813, starting with the second figure from right:

$$\overline{3} \quad 8 \quad \overline{1} \quad 3$$

We then osculate as normal except the *'any carry figure is counted as negative'*.

$$\overline{3} \quad 8 \quad \overline{1} \quad 3$$
$$0 \quad 32 \quad 8$$

$3 \times 3 + \overline{1} = 8$

$8 \times 3 + 8 = 31,$

$2 \times 3 + \overline{3} + \overline{3} = 0$; this zero indicates that 3813 is divisible by 31.

- Test whether 367164 and 6454 divisible by 7

First of all multiply 7 by 3 which give 21, so the negative osculator is 2.

For **367164**

$$\overline{3} \quad 6 \quad \overline{7} \quad 1 \quad \overline{6} \quad 4$$
$$0 \quad 12 \quad 3 \quad 5 \quad 2$$

For **6454**

$$\overline{6} \quad 4 \quad \overline{5} \quad 4$$
$$\overline{7} \quad 10 \quad 3$$

Since $\overline{7}$ is a multiple of 7 we find that both these numbers are divisible by 7.

- Is 11594 divisible by 62

As 62 = 31 x 2 we need to test for 31 and 2.

The number is clearly divisible by 2 so we test for 31 by osculating with 3.

$$1 \quad \bar{1} \quad 5 \quad \bar{9} \quad 4$$
$$0 \quad 10 \quad 14 \quad 3$$

The number is divisible by both 2 and 31 and therefore divisible by 62.

Exercise

Test the divisibility

6039 by 91   2914 by 31   9576 by 21   20022 by 71   73472 by 41   63909 by 81   1728 by 91

7071 by 17   14715 by 27   7071 by 61   178467 by 31   45787 by 7   14715 by 27   7071 by 61

178467 by 31   45787 by 7   2394 by 42   4838 by 82   17949 by 93   9658 by 11

Osculating with groups of digits

For 299, the positive osculator (PO) is 3 and we osculate with pairs of digits.

Similarly for 60001 we use negative osculator (NO) 6, with the groups of four digits.

Example
- Is 80132 divisible by 299

Here we have PO as 3 so: 8   01   32
                           299 97

We split the given number into pairs of digits starting from right, and then osculate with 3. So $32 \times 3 + 01 = 97$
Then $97 \times 3 + 8 = 2999$
Therefore, 80132 is divisible by 299.

- Is 1625325 divisible by 5001

Here we have NO as 5 so: 1 $\overline{625}$ 325
                       0   1000
Split into groups of three.
$325 \times 5 - 625 = 1000$.

The 1 in 1000, count as $\bar{1}$ as it is to be carried, so $000 \times 5 + 1 + \bar{1} = 0$.
So 165325 is divisible by 5001.

- Is 358211 is divisible by 701

Here we have NO as 7, so: 35 $\overline{82}$ 11

0 $\overline{5}$

11 x 7 – 82 = $\overline{5}$.

$\overline{5}$ x 7 + 35 = 0.

So 358211 is divisible by 701.

Exercise

Test the following divisibility

647546 by 199   3304662 by 299   2784607 by 1999

264385741 by 1999   1821849 by 499   12256985 by 701

41221372 by 1001   359061579 by 999

Chapter -4

Division

DIVISION

Basics

The number that is divided is called the dividend.

The number that divides the dividend is called divisor.

The number of times the dividend is divisible by the divisor is called the quotient.

If the dividend is completely not divisible by the divisor it leaves behind a remainder.

Example: How do you check if the division is right?

$$8 \div 2 = 4$$

quotient

dividend   divisor

$$4 \longrightarrow \text{quotient}$$

divisor $\longleftarrow 2 \overline{)8} \longrightarrow$ dividend

Dividend	=	Quotient	x	Divisor	+	Remainder
Here		the		remainder		is		0
Hence		8	=	4	x	2

Division by All from 9 and last from 10

This method of division is useful when every digit of divisor is greater than 5.

Take a base (in the power of 10) nearest to the divisor and write its complement below the original divisor. Complement = Base – Divisor

The number of the digits to be placed in the remainder should be equal to the number of zeroes in the base and is separated from the dividend by a stroke.

The first digit of the dividend is brought straight down and it gives the first digit of the quotient.

The first quotient digit is multiplied by the complement, and placed below the next dividend digit.

The digits in the second column are added up and the answer is the next dividend digit.

The new quotient digit is multiplied by the complement, and placed below the next dividend digit.

Repeat the process until the number in the remainder column is less than that of the original divisor.

For a single digit divisor a space of one line is left before drawing the answer line and for two digits divisor, the space of three lines are left and so on.

111 ÷ 8

	Divis	Divid	Rem
	Col	Col	Col

Base - 10

```
8 | 1 1 / 1
2 |   2   6
  |_____
    1 3 / 7
```

Answer is **13**

remainder **7**

Explanation

For a single digit divisor a space of one line is left before drawing the answer line. A remainder stroke is placed so that the number of digits in the divisor, one is the same as the number of digits after the stroke.

The complement of the divisor, 2 is written below 8.

The first digit of the dividend is brought straight down into the answer which is 1 in this case.

The first quotient digit,1 is multiplied by the complement, 2 and the answer (1 x 2 =2) is placed below 1

The second column of dividend is added up, 1 + 2= 3, and the answer is the next quotient digit.

Multiply the second digit of quotient with the complement (3 x 2= 6), and place it in the remainder column, next to the second digit of the dividend which is 1

Add up the digits of the remainder column, 1 + 6 =7 and 7 < 8 (divisor) so the process stops. So the answer is 13 remainder 7.

10025 ÷ 88

	Divis	Divid Rem
	Col	Col Col

Base - 100

```
        88 |1 0 0/ 2 5
        12 | 1 2
           |   1  2
           |_____3 6
            1 1 3 / 8 1
```

Answer is **113**
remainder **81**

Explanation

For a two digits divisor three lines spacing is left before drawing the answer line. A remainder stroke is placed so that the number of digits in the divisor, two is the same as the number of digits after the stroke.

The complement of the divisor, 12 is written below 88.

The first digit of the dividend is brought straight down into the answer which is 1 in this case.

The first quotient digit, 1 is multiplied by the complement, 12 and the answer (1 x 12 = 12) is placed below 0

The second column of dividend is added up, 0 + 1= 1, and the answer is the next quotient digit.

Multiply the second digit of quotient with the complement

(1 x 12= 12), and place it in the dividend column, next to the second digit of the dividend which is 0.

The third column of dividend is added up, 0 + 2 + 1= 3, and the answer is the next quotient digit.

Multiply the third digit of quotient with the complement (3 x 12= 36), and place it in the remainder column, next to the third digit of the dividend which is 2.

Add up the digits of the remainder column, 25 + 2 + 36 = 81 and 81 < 88 (divisor) so the process stops. So the answer is 113 remainder 81.

1374 ÷ 878

Base - 1000

```
878 |1 / 3 7 4
122 |   1 2 2
      1 / 4 9 6
```
Answer is **1**
remainder **496**

11111111÷99979

Base - 100000

```
99979 |1 1 1 / 1 1 1 1 1
00021 | 0 0   0 2 1
       |  0   0 0 2 1
       |      0 0 0 2 1

        1 1 1   1 3 4 4 2
```
Answer is **111**
remainder **13442**

1108 by 79

We set out the sum marking off two figures on the right (as we have a 2-figure divisor) and leave two rows as there are to be two answer figures.

90

$$79)11\ |\ 08$$
$$21\quad 2\ |\ 1$$
$$\underline{\qquad 6\ 3}$$
$$13\ |\ 8\ 1 \qquad = 14\ \text{remainder}\ 2$$

1121123 by 8989

$$8989)11\ 2\ |\ 1123$$
$$1011\quad 10\ |\ 11$$
$$2\ |\ 022$$
$$4044$$
$$\underline{\qquad\qquad}$$
$$12\ 4\ |\ 64\ 87$$

Divide the following?

89)1021   88)1122   79)1001   88)2111   97)1111

888)10011   887)11243 899)21212   988)30125   8899)201020

88)10025   99979)11111111

When the remainder exceeds the divisor

When there is a remainder which is greater than divisor then we simply add the complement to the remainder and if number of digits in the remainder is more than the noughts in the base of the divisor then the excess digit is carried over to the left.

4261 ÷ 9

$$
\begin{array}{r}
9|4\ 2\ 6\ /\ 1 \\
1|\ \ 4\ 6\ \ \ 12 \\
\hline
4\ 6\ 12/\ 13 \\
+1 \\
\hline
4\ 7\ 2\ \ \ \ 14 \\
\\
4\ 7\ 3\ \ \ 4
\end{array}
$$

Answer is **473**

remainder **4**

Here in the third column we have 12, so 1 of 12 is carried to the left to make it 4 72.

In remainder column 13 is greater than the divisor 9 so we add the complement 1 to the remainder to get 14. 1 of 14 is carried to the left to make 473 and 4 is left as the remainder.

1294567 ÷ 89997

Base - 100000

(R = 124606 > 89997)

```
89997 | 12 / 9 4 5 6 7
10003 | 1    0 0 0 3
      |      3 0 0 0 9
      ─────────────────
       1 3 / 1 2 4 6 0 6
          + 1 0 0 0 3
      ─────────────────
       1 3 1 3 4 6 0 9
            ⌣
      ─────────────────
          14    3 4 6 0 9
```

Answer is **14**
remainder **34609**

11199171 ÷ 99979

Base − 100000

(R = 101502 > 99979)

```
99979 | 111 / 9 9 1 7 1
00021 | 0 0    0 2 1
      |   0    0 0 2 1
      |        0 0 0 2 1
      ──────────────────
       1 1 1 / 1 0 1 5 0 2
            + 0 0 0 2 1
      ──────────────────
       1 1 1    1 0 1 5 2 3
                  ⌣
      ──────────────────
          1 1 2    0 1 5 2 3
```

Answer is **112**
remainder **01523**

Division by Transpose and Adjust

This method of division is used when the divisor is a little more than the base (10 or power of 10). Here we transpose all the numbers of divisor greater than the base into vinculum number. It works effectively when the first digit of the divisor is 1.

In the divisor column, write the complement of the divisor from the base with changed sign below the original divisor.

Multiplication of quotient digits is performed with the transposed (changed sign) complement for each digit.

256 ÷ 11

$$11 | 2\ 5 / 6$$
$$\bar{1} |\ \ -2\ \ -3$$
$$\overline{\qquad\qquad}$$
$$2\ 3 /\ 3$$

Answer is **23**
Remainder **3**

Explanation

The changed sign complement of the divisor, -1 or $\bar{1}$ is written below 11.

The first digit of the dividend is brought straight down into the answer which is 2 in this case.

The first quotient digit, 2 is multiplied by the complement, $\bar{1}$ and the

94

answer -2 is placed below 5

The second column of dividend is added up, 5 + (-2) = 3, and the answer is the next quotient digit.

Multiply the second digit of quotient with the complement $(3 \times \bar{1} = -3)$, and place it in the remainder column, next to the second digit of the dividend which is 6

Add up the digits of the remainder column, 6+ (-3) =3 and 3 < 11 (divisor) so the process stops. So the answer is 23 remainder 3.

23689 ÷ 112

```
112 | 2 3 6/ 8 9
 __
 12 | -2-4
    |  -1 -2
    |____-1-2_____
      2 1 1 /5 7
```

Answer is **211**
remainder **57**

395166 ÷ 1321

```
1321 | 3 9 5 / 1 6 6
 ___
 321 |  -9-6  -3
     |  0  0 0
     |_____3 2 1____
       3 0-1 / 1 8 7
```

$30\bar{1}$ / 187
299/187

Answer is **299**
remainder **187**

Divide the following?

123)1377   131)1481   121)256   132)1366   1212)13545
161)1781   1003)321987
111)79999

Straight Division (also known as Flag method)

This is a general method for dividing any number. It can be applied to divisors and dividends of any size.

When the divisor is written down all its digits, except for the first are hoisted into the flag position.

The dividing is then only done by the first digit.

There is one flag digit for two-digit divisors and two flag digits for three-digit divisors and so on.

The number of flag digits indicates the number of digits which need to be placed after the remainder stroke.

337 ÷ 24

```
        4
   2  | 3 3 / 7
      | 1   1
      _____
      1 4 / 1
```

Answer is **14**
remainder **1**

Explanation

We take only the first digit i.e. 2 out of the divisor 24 in the divisor column and put the other digit i.e. 4 on the top of the flag. The number of flag digits indicates the number of digits which need to be placed after the remainder stroke.

Here the main divisor is 2 so the entire division is to be carried out by 2. On dividing 3 by 2 we get 1 as the quotient and 1 remainder. Remainder 1 is placed below just before the second digit 3 making 13 as the second dividend.

Subtract the product of flag digit (4) and the first quotient (1) from 13. This is 13-(4x1) = 9, and 9 is then divided by 2 which gives 4 rem 1. The 4 is the next quotient and the remainder is prefixed to the 7 making it 17.

In remainder portion we do not divide but simply subtract the product of the previous quotient digit (4) and the flag digit(4) from the remainder(17). This is 17 − (4x4)= 1 and this is the remainder.

The answer is 14 remainder 1.

38982 ÷ 73

$$
\begin{array}{r}
3 \\
7\,\big|\,3\;8\;9\;8\,/\,2 \\
\underline{3\;3\quad 1} \\
5\;3\;4\,/\,0
\end{array}
$$

Answer is **534**
remainder **0**

Explanation

The remainder stroke is placed so that the number of digits to the right of it is the same as the number of flag digits.

Main divisor = 7 , Flag = 3

As the first digit from the left of the dividend (3) is less than 7, we take 38 as our first dividend. Divide 38 by 7 which gives 5 remainder 3. Next dividend =39.

New Dividend = 39 – (3x5) = 24

24 divided by 7 = 3 rem 3. next dividend = 38

New dividend = 38 – (3x3) = 29

29 divided by 7 = 4 rem 1. next dividend = 12

Remainder = 12 – (3x4) =0.

The answer is 534 and remainder 0.

Altered Remainders

In straight division method whenever we encounter a situation where the new dividend in the middle is coming out to be zero or negative (after the subtraction of dividend from the product of quotient and flag) or the remainder is coming out to be zero (after dividing new dividend by the divisor) then we subtract 1 from the quotient and find the remainder. For example

Divide 31 by 3 = 10 remainder 1

31 / 3 = 10 rem 1 or 9 rem 4 or 8 rem 7 or 7 rem 10 and so on are the altered remainders for 31/3. By repeatedly

subtracting 1 from the quotient, 10, we find the next quotients, 9, 8 & 7.

With each subtraction of 1 we add the divisor 3 to the remainder, i.e. 1+ 3 = 4, 4 + 3 = 7 and 7 + 3 = 10

When flag has more digits than main divisor (Altered Remainder)

7238761 ÷ 524

(Here remainder is zero so we follow Altered remainder method)

```
  24
5 | 7 2 3 8 7 / 6 1
      2
    1
```

```
    24
 5 | 7 2 3 8 7 / 6 1
      2 5
      1 3
```

```
  24
5 | 7 2 3 8 7 / 6 1
    2 5 3
    1 3 8
```

```
    24
 5 | 7 2 3 8 7 / 6 1
     2 5 3 5
       1 3 8 1
```

```
  24
5 | 7 2 3 8 7 / 6 1
    2 5 3 5   3
    1 3 8 1 4
```

Answer is **13814**

Remainder **225**

Here we take two digits as flag, so remainder column will have two digits.

Main divisor = 5 , Flag = 24

99

On dividing 7 by 5 we get 1 rem 2, next dividend = 22
22 −(1x2)= 20, 20 divided by 5 gives 4 rem 0. Since rem=0 so we have to take quotient below 4. Revised quotient = 4-1= 3. This is because we cannot consider the remainder zero in the middle of division. The same process will be applicable in case the new dividend in the middle is either zero or negative.

Now revised quotient is 3 and rem =5. Next dividend = 53.

New dividend = 53 - (sum of cross product of two flag digits and two quotients)= 53 − (2x3 + 1x4) = 43

43 divided by 5 = 8 rem 3, next dividend = 38.

New dividend 38 − (sum of cross product of 38 and 24) = 38 - (3x4 + 2x8) = 38 - 28 =10.

10 divided by 5= 2 rem 0 (use altered remainder).

Revised quotient = 2-1 =1 and rem = 5. Next dividend = 57.

New dividend = 57 − (sum of cross product of 24 and 81) = 57 − (2x1 + 4x8) = 57- 34= 23.

23 divided by 2 = 4 rem 3.

Remainder = 361 − (sum of cross product of 24 and 14) x 10 − (last flag digit x last quotient) =

361 − (2x4 + 4x1) x 10 − (4x 4) = 361 −(8+4) x 10 − 16 = 361 − 120 − 16 = 225.

Hence the answer is 13814 and remainder is 225.

13579 ÷ 975

```
  75              75                75
9│1 3 5 / 79    9│1 3 5 / 79      9│1 3 5 / 79
 └─────────      │ 4               │ 4  11
   1             └─────────        └─────────
                  1 4                1 3
```

Answer is **13** remainder **904**

Explanation

Divide 13 by 9 = 1 rem 4, next dividend = 45

New dividend = 45 – 1x7 = 38

Divide 38 by 9 = 4 rem 2

Remainder = 279- (cross product of 75 and 14) x 10 – (product of last flag and last quotient) =

279 – (7x4 + 5x1) x 10 – (5x4) = 279 – 33 x 10 – 20 = 279 – 330 – 20 = -71 (value in negative so altered remainder method is followed)

Revised quotient = 4-1 =3 rem 11

New remainder = 1129 – (3x7 + 1x5) x10 – (3x5)= 1129 – 260 -15 = 904.

Hence the answer is 13 remainder 904

Divide the following?

62)19902	73)44749	88)1936	46)360293	73)511717
56)301291	53) 12233	71) 9018		
72) 8910	112) 23658	76) 22222	82)651258	

Division by Nine (Short cut)

71 ÷ 9

First number of the dividend is the Quotient and the sum of the

numbers is remainder.

In the above example

Quotient = 7, Remainder = 8

134 ÷ 9 = 14 remainder 8

The answer consists of 1, 4 and 8.
1 is just the first figure of 134.
4 is the total of the first two figures 1+ 3 = 4, and 8 is the total of
all three figures 1+ 3 + 4 = 8.
842 ÷ 9 = 8$_1$2 remainder 14 = <u>92 remainder 14</u>

Actually a remainder of 9 or more is not usually permitted
because we are trying to find how
many 9's there are in 842.

Since the remainder, 14 has one more 9 with 5 left over the final
answer will be 93 remainder 5

Divide the following numbers by 9

51	34	71	44	60	71	46	34

Now let's extend the same method for the division of larger numbers

2311 ÷ 9

The first figure of the dividend is the first figure of the answer (Quotient)

Each figure of the dividend is added to the next figure in the dividend to give next figure of the answer

The last number we write down is the remainder

9) 2 3 1 1
   2 5 6 7

256 is the quotient and 7 is the remainder

11202 ÷ 9

9) 1 1 2 0 2
   1 2 4 4 6

1244 is the quotient and 6 is the remainder

3172 ÷ 9

9)3 1 7 2
   3 4 11 13

In this example 1 in 11 must be carried over to the 4, giving 351, and there is also another 1 in the remainder so we get Quotient **352 and remainder 4**

21.2 ÷ 9

9) 21.2 0 0 0

2.3555

Here we have a decimal point and the answer is given as decimal without a remainder. Adding 5 to 0 repeatedly, gives the recurring 5

Divide the following numbers by 9?

1234	311101	203010	444	3272	5747	8252
3102	20002	444	713	443322	75057	12345

Division by Eight

The first figure of the dividend is the first figure of the answer (Quotient)

Twice of each figure of the dividend is added to the next figure in the dividend to give next figure of the answer

The last number we write down is the remainder

1101 ÷ 8

8)1 1 0 1

   1 3 6 13

   13 7 5

137 is the quotient and 5 is the remainder

Divide the following numbers by 8?

101	1101	2121	11111	132	111	3411

Division by 11, 12, 13 etc

In case of 11, 12 and 13 we subtract the last digit at each step, instead of adding it

Example: 3411 ÷ 11

11) 3 4 1 1
    3 1 0 1

We bring down the first digit of dividend (3) then followed the following steps

   4-3=1

   1-1=0

   1-0=1

The answer is 310 and the remainder is 1

Example: 523 ÷ 11

$11\overline{)5\;2\;3}$

$5\;\overline{3}\;6$ =47 remainder 6

3411 2

$12\overline{)3\;4\;\;1\;1}$

$3\;\overline{2}\;\overline{5}\;9$

The answer is 284 and remainder 3

Divide the following?

543 ÷ 11   20304 ÷ 11   81726 ÷ 11   345 ÷ 11   1489 ÷ 12   333
÷ 12   5151 ÷ 12 9184 ÷ 12

When the divisor is close to the base

235 ÷ 88

We have to know how many times 88 can be taken from 235 and what the remainder is.

Since every 100 must contain an 88 there are clearly 2 88's in 235 and the remainder will be 2 12's (as 88 is 12

106

short of 100) plus the 35 in 235. So the answer is 2 and the remainder = 24+35 = 59

Chapter -5

Square of a number is obtained when you multiply a number with itself. You might have used it for solving problems that involve calculating the surface area of squares, cubes and other geometrical figures. Vedic math provide very powerful techniques to find out squares of multi digit numbers very fast. Some of these techniques are discussed in the section below

Squares

One more than the previous one Method

This method for squaring is valid only if a number ends with 5.

- Square the unit digit.
- Multiply the remaining digit by number one more than itself.

Square of a number ending with 5 = Left digit(s) x {Left digit(s) + 1} | 25

Example:-

Find the square of **'35'**

Square of unit digit = $(5)^2$ = 25

Multiply 3 with (3 + 1 = 4) = 3 x 4 = 12

So, $(35)^2$ = 1225

Exercise

Find the square of the following

25 45 55 65 75 85 95   15   105

Proportionately Method

This method is useful for squaring a two digit number.

- It consists of three columns.
- The 1st column consists of squaring of digit placed at tens.
- The 2nd column contains the product of unit digit and tens digit. Double this figure.
- The 3rd column contains the square of the unit digit.
- Now arrange the answer by adding from right to left keeping only one digit in 3rd and 2nd column.

Example:-Find the square of square of **'67'**

Column 1st   Column 2nd   Column 3rd

$(6)^2 = 36$ 6 x 7 = 42 $(7)^2 = 49$

2 x 42 = 84

36 | 84 | 49

+   +

36 + 8 | 4 + 4 | 9 = 4489

So, $(67)^2 = 4489$

Find the square of **'86'**

Column 1st	Column 2nd	Column 3rd

$(8)^2 = 64$ $8 \times 6 = 48$ $(6)^2 = 36$

$2 \times 48 = 96$

64 | 96 | 36

         +   +

$64 + 9 \mid 6 + 3 \mid 6 = 7396$

So, $(86)^2 = 7396$

Find the square of **'98'**

Column 1st	Column 2nd	Column 3rd

$(9)^2 = 81$ 　　　　　 $9 \times 8 = 72$ 　　　 $(8)^2 = 64$

　　　　　　　　　　　　 $2 \times 72 = 144$

81 　　|　　 144 　　|　 64

　　　+　　　　　　　　　　+

$81 + 14 \mid 4 + 6 \mid 4 = 9604$

So, $(98)^2 = 9604$

Near Base Method

This method works when the number to be squared is nearer to the base, which is in the power of 10 (10, 100, 1000....) or the multiple of 10 (20, 30, 40, 50....).

- Find the excess or the deficiency of a number to be squared with respect to its base. The excess or deficiency is called as "Deviation".
- Set up the square of the deviation at the end.

Power of 10 Base (10, 100, 1000....)

Example:-

Find the square of '**12**'

The nearest base of 12 is 10.

Deviation from the base = 12 – 10 = 2

Add 2 to 12 and the set up the square of 2 at the end.

$$= 12 + 2 \mid (2)^2 = 144$$

So, $(12)^2 = 144$

Find the square of '**16**'

The nearest base of 16 is 10.

Deviation from the base = 16 – 10 = 6

Add 6 to 16 and the set up the square of 6 at the end.

$= 16 + 6 \mid (6)^2 = 22 \mid 36 = 256$

$\underbrace{\qquad}$
$+$

So, $(16)^2 = 256$

Find the square of '**91**'

The nearest base of 91 is 100.

Deviation from the base $= 91 - 100 = -9$

Subtract 9 from 91and the set up the square of (-9) at the end.

$= 91\text{-}9 \mid (-9)^2 = 82 \mid 81 = 8281$

So, $(91)^2 = 8281$

Find the square of '**108**'

The nearest base of 108 is 100.

Deviation from the base = 108 − 100 = 8

Add 8 to 108 and the set up the square of 8 at the end.

= 108 + 8 | $(8)^2$ = 116 | 64 = 11664

So, $(108)^2$ = 11664

Find the square of '**997**'

The nearest base of 997 is 1000.

Deviation from the base = 997 − 1000 = -003

Subtract 3 from 997and the set up the square of (-003) at the end.

= 997-3 | $(-003)^2$ = 994 | 009 = 994009

So, $(997)^2$ = 994009

Multiple of 10 Base (20, 30, 40, 50….).

Example:-

Find the square of '**32**'

As 32 is not nearer to base 10 so 30 (3x10) is taken as working base.

Deviation from working base = 32 – 30 = 2

Add 2 to 32 and multiply this figure by 3 and set up the square of 2 at the end.

= 3 x (32 + 2) | $(2)^2$ = 3 x 34 | 4 = 102 | 4 = 1024

So, $(32)^2$ = 1024

Find the square of '**48**'

As 48 is not nearer to base 10 so 50 (5x10) is taken as working base.

Deviation from working base = 48 – 50 = -2

Subtract 2 from 48 and multiply this figure by 5 and set up the square of -2 at the end.

= 5 x (48 - 2) | $(-2)^2$ = 5 x 46 | 4 = 230 | 4 = 2304

116

So, $(48)^2 = 230$

Find the square of **'64'**

As 64 is not nearer to base 10 so 60 (6x10) is taken as working base.

Deviation from working base = 64 – 60 = 4

Add 4 to 64 and multiply this figure by 6 and set up the square of 4 at the end.

= 6 x (64 + 4) | $(4)^2$ = 6 x 68 | 16 = 408 | 16 = 4096

So, $(64)^2 = 4096$

Exercise

Find the square of the following
14   18   96   41   59   77   92   42   63   29   33   102 999   993   1004   1013

Duplex Method

This method is one of the best squaring method in Vedic Math because of its universal application. The group formation in the Duplex method is same as done in "Dot & Stick "method in multiplication.

Duplex of a number

Duplex of 1 digit number = Square of that number.

- Duplex of 4 = 4^2 = 16

Duplex of 2 digit number = 2 x (Product of the digits)

- Duplex of 24 = 2 x (2x4) =16

Duplex of 3 digit number = 2 x (Product of 1st digit and 3rd digit) + square of middle digit

- Duplex of 256 = 2 x (2x6) + 5^2 = 24 + 25 = 49

Duplex of 4 digit number = 2 x (Product of 1st digit and 4th digit) + 2 x (Product of 2nd digit and 3rd digit)

- Duplex of 5678 = 2 x (5x8) + 2 x (6x7) = 164

Duplex of 5 digit number = 2 x (Product of 1st digit and 5th digit) + 2 x (Product of 2nd digit and 4th digit) + square of middle digit

- Duplex of 56789 = 2 x (5x9) + 2 x (6x8) + 7^2 = 235

Group Formation of a number

The grouping of $(24)^2$ will follow the pattern of 121 of $(11)^2$.

The groups of 24 are:-

2	24	4
1digit	2digit	1digit

The grouping of $(245)^2$ will follow the pattern of 12321 of $(111)^2$.

The groups of 245 are:-

2	24	245	45	5
1digit	2digit	3digit	2digit	1digit

The grouping of $(2456)^2$ will follow the pattern of 1234321 of $(1111)^2$.

The groups of 2456 are:-

2	24	245	2456	456	56	6
1dgt	2dgt	3dgt	4dgt	3dgt	2dgt	1dgt

How does Duplex method work?

- Form the group of the number to be squares as shown above.
- Write the duplex of each group digit.
- Once the duplex value for each group is written, add the figures from right to left, as done in multiplication.

Examples:-

Find the square of **'24'**

The groups for 24 are

= 2	24	4
2^2	2 x 2 x 4	4^2 (Duplex)

= 4 | 16 |16 = 576

+   +

So, $(24)^2 = 576$

120

Find the square of '**75**'

The groups for 75 are

= 7   75   5

7^2 2 x 7 x 5 5^2 (Duplex)

= 49 | 70 | 25 = 5625

  ‿   ‿

  +   +

So, $(75)^2 = 5625$

Find the square of '**245**'

The groups for 245 are

= 2   24   245   45   5

2^2 2 x 2 x 4   2x2x5+4^2 2x4x5 5^2

          (Duplex)

= 4 | 16 | 36 | 40 | 25 = 60025

  ‿   ‿   ‿   ‿

  +   +   +   +

So, $(245)^2 = 60025$

Find the square of '**268**'

The groups for 268 are

= 2   26   268   68   8

2^2 2 x 2 x 6   2x2x8+ 6^2 2x6x8 8^2
      (Duplex)

= 4 | 24 | 68 | 96 | 64 = 71824
   +   +   +   +

So, $(268)^2 = 71824$

Find the square of '**1234**'

The groups for 1234 are

= 1   12   123   1234   234   34   4

1^2 2 x 1 x 2   2x1x3+ 2^2 2x1x4+2x2x3   2x2x4+3^2 2x3x4
     4^2 (Duplex)

122

= 1| 4 | 10 | 20 | 25 | 24 | 16 = 1522756

So, $(1234)^2$ = 1522756

Find the square of 23456

The groups for 23456 are

= 2   23   234 2345   23456   3456   456 56   6

Duplex of the above groups

$2 = 2^2 = 4$

$23 = 2 \times 2 \times 3 = 12$

$234 = 2 \times 2 \times 5 + 3^2 = 25$

$2345 = 2 \times 2 \times 5 + 2 \times 3 \times 4 = 44$

$23456 = 2 \times 2 \times 6 + 2 \times 3 \times 5 + 4^2 = 70$

$3456 = 2 \times 3 \times 6 + 2 \times 3 \times 5 = 76$

$456 = 2 \times 4 \times 6 + 5^2 = 73$

$56 = 2 \times 5 \times 6 = 60$

$6 = 6^2 = 36$

Arrange the duplex

4 | 12 | 25 | 44 | 70 | 76 | 73 } 60 | 36 = 550183936

So, $(23456)^2 = 55018396$

Exercise

Find the square of the following

29   34   47   79   97   109   99   168   192   2356
6254

2591   12345   41326

Square root

The square root of a number x is that number which when multiplied by itself, gives x as the product.

We write the square of a number x as $\sqrt{x}$, for example $\sqrt{4}$ =2, $\sqrt{9}$ = 3, $\sqrt{16}$ =4.

Important points to remember

- If the unit digit of any number is 2, 3, 7, 8, it can't be a perfect square root.
- If a number ends in an odd number of zeros, it can't be a perfect square root.
- Square numbers only have digit sums of 1, 4, 7, 9 and they only end in 0, 1, 4, 5, 6, 9.

Unit digit of a square	Unit digit of a square root
0	0
1	1 or 9
4	2 or 8
5	5
6	4 or 6
9	3 or 7

In the above table:

- A square ending in 1 must have either 1 or 9, as the last digit of its square root.
- A square can end in 4, only if the square root ends in 2 or 8.
- A square ends with 5; its square root too ends with 5.
- A square ending in 6 must have 4 or 6 as the last digit in its square root.
- A square ending in 9 must have 3 or 7 as the last digit in its square root.

So, it's easy to tell whether a number is perfect square or not.

Square Root of Perfect Square

Example

- Square root of $\sqrt{6889}$

First of all make two groups of figures starting from right, 68'89, so we expect a 2- figure answer. Now, looking at 68 we see that 68 is greater than 64 which is $(8)^2$ and less than 81 which is $(9)^2$ so the figure must be 8.

Or looking at another way 6889 is between 6400 and 8100

$$6400 = 80^2$$

$$6889 = 8?^2$$

$$8100 = 90^2$$

So $\sqrt{6889}$ must be between 80 & 90. Now we look at the last figure of 6889, which is 9.

Any number ending with 3 or 7 will end with 9 when it is squared so the number could be 83 or 87.

There is two easy ways of deciding. One is to use the digit sums.

If $87^2 = 6889$ then converting to digit sums we get $6^2 = 4$, which is not correct. But $83^2 = 6889$ becomes $4 = 4$, so the answer must be 83.

The other method is to recall that since 85^2 = 7225 and 6889 is below this $\sqrt{6889}$ must be below 85. So it must be 83.

- Square root of $\sqrt{5776}$

The 57 is between 49 and 64, so the first figure must be 7.

The 6 at the end tells us the square root ends with 4 or 6. so the answer is 74 or 76.

74^2 = 5776, digit sums for this is 4 =7 which is not true in terms of digit sums, so 74 is not the answer.

76^2 = 5776 becomes 7 = 7 so 76 is the answer.

Alternatively to choose between 74 and 76 we note 75^2 = 5625 and 5776 is greater than this so the square root must be greater than 75. So it must be 76.

Exercise

2116	5329	1444	6724	3481	4489	8836	361
784	3721	2209	4225	9604	5929		

- Square root of $\sqrt{31329}$

Let's first mark off the pairs of digits from the right here we get 3'13'29. We have three groups, indicating that the answer is a 3- figure number. However if you know all the square numbers up to 20^2 we can get the answer by this method. Now make the pairs of digits as 313'29.

Since 313 lies between 289 (17^2) and 324 (18^2) the first two figures must be 17 and the last figure is 3 or 7, so 173 and 177 are two possibilities.

The digit sum will then confirm 177 as the right one.

Exercise

26896	32761	16129	24964	36864	18496	21025
29929	14161	11236				

Methods for finding out the square root of a number

Prime Factorization Method

When a given number is a perfect square, we find its square root by the following steps:

- Find the prime factors of the number
- Make pairs of the same factor
- Take the product of the prime factors, choosing one factor out of every pair.

Example

- Find the square root of 324.

By Prime Factorization, we get 324 = 2x2 x 3x3 x 3x 3

$$\sqrt{324} = 2 \times 3 \times 3 = 18$$

2	3 2 4
2	1 6 2
3	8 1
3	2 7
3	9

130

- Find the smallest number by which 396 must be multiplied so that the product becomes a perfect square.

```
2 | 3 9 6
  |_____
  | 1 9 8
  |_____
3 | 9 9
  |_____
3 | 3 3
  |_____
```

By Prime Factorization, we get 396 = 2x2 x 3x3 x 11

It is clear that in order to get a perfect square, we must multiply 396 by one more 11.

Division Method

- Starting with the digit in the units place, group the digits in pair. Each pair and the remaining digit (if any) is called a period.
- Think of the largest number whose square is equal to or less than the first period. Take this number as the Divisor and also the Quotient.
- Subtract the product of the Divisor and the Quotient from the first period and bring down the next period to the right of the Remainder. This becomes the new Dividend.
- Now, the new Divisor is obtained by taking two times the Quotient and annexing with it a suitable digit which is also taken as the next digit of the Quotient,

131

chosen in such a way that the product of the new Divisor and this digit is equal to or just less than the new Dividend.

- Repeat steps (2), (3) and (4) till all the period have been taken up.

Now the Quotient so obtained is the required square root of the given number.**Example:**

- Square root of 16384

$\sqrt{16384}$

$$
\begin{array}{r|l}
1 & \overline{1}\ \overline{63}\ \overline{84}\ (\ 128 \\
 & 1 \\
\hline
22 & 63 \\
23 & \\
\hline
 & 44
\end{array}
$$

- Square root of 10609

$\sqrt{10609}$

$$
\begin{array}{r|l}
1 & \overline{1}\ \overline{06}\ \overline{09}\ (\ 103 \\
 & 1 \\
\hline
203 & 0609
\end{array}
$$

- Find the greatest number of four digit which is a perfect square

The greatest four digit number is 9999 and the square root of this number is

$$
\begin{array}{r|l}
9 & \overline{99}\ \overline{99}\ (\ 99 \\
\hline
& 81 \\
\hline
189 & 1899 \\
\end{array}
$$

This shows that $(99)^2$ is less than 9999 by 198. Hence the required number is (9999 - 198) = 9801

- Find the smallest number of six digit which is a perfect square.

The smallest number of six digit is 100000, which is not a perfect square. Now, we must find the smallest number which when added to 100000 gives a perfect square. This perfect square is the required number.

Let's find the square root of 100000.

```
      |  __ __ __
   3  | 10 00 00 ( 316
      |_____
      |
      | 9
      |_____
  61  | 100
      |_____
      |
      |    61
```

So, $(316)^2 < 100000 < (317)^2$

Therefore, the smallest number to be added is $(317)^2$ - 100000 = 489

Hence the required number is 100000 + 489 = 100489

Square roots of the numbers in decimal form

Example:

- Find $\sqrt{42.25}$

$$
\begin{array}{r|l}
6 & \overline{42}.\overline{25}\,(6.5 \\
\hline
 & 36 \\
\hline
125 & 625 \\
\end{array}
$$

Therefore, $\sqrt{42.25}$ = 6.5

- Find $\sqrt{1.96}$

$$
\begin{array}{r|l}
1 & \overline{1}.\overline{96}\,(1.4 \\
\hline
 & 1 \\
\hline
24 & 96 \\
\end{array}
$$

Therefore $\sqrt{1.96}$ = 1.4

Square root correct up to certain places of decimal

Example

- Find $\sqrt{2}$ correct up to two places of decimal

```
         ___ __ __ __
  1 | 2 . 00 00 00  (1.414
    |
    | 1
    |_____
 24 | 100
    |
    |  96
    |_____
281 | 400
```

Therefore, $\sqrt{2}$ = 1.41

- Find $\sqrt{0.8}$ correct up to two places of decimal

```
         __ __ __
  1 | 0. 80 00 00  (0.894
    |_____
    | 64
    |_____
169 | 1600
    |_____
    | 1521
```

Therefore, $\sqrt{0.8}$ = 0.894

136

Chapter -6

Cube of a number is obtained by multiplying a number by itself three times. We deal with the cubes of number while solving problems related volume of cube , spheres and other geometrical figure and sometimes while dealing with problems related to commercial mathematics. Vedic mathematics provide lot techniques that helps in finding out the cube of multi digit numbers mentally and faster than conventional methods

Cubes

E. g. $3^3 = 3 \times 3 \times 3 = 27,$

$2^3 = 2 \times 2 \times 2 = 8,$

$4^3 = 4 \times 4 \times 4 = 64,$ and

$5^3 = 5 \times 5 \times 5 = 125$

Algebraic Method to find cube

$(a + b)^3 = a^3 + 3 a^2 b + 3 a b^2 + b^3$ and

$(a - b)^3 = a^3 - 3 a^2 b + 3 a b^2 - b^3$

Below are some examples on how these formulas work

$(21)^3 = (20 + 1)^3 = (20)^3 + 3 \times (20)^2 \times 1 + 3 \times (1)^2 \times 20 + (1)^3$

   $= 8000 + 1200 + 60 + 1$

   $= 9261$

$(21)^3 = (30 - 9)^3 = (30)^3 - 3 (30)^2 \times 9 + 3 \times (9)^2 \times 30 - (9)^3$

138

$= 27000 - 24300 + 7290 - 729$

$= 9261$

In Vedic mathematics there are several sutras that help in finding the cubes of a number faster than the normal method. In the following sections we will discuss all of them

Method 1 - Anurupyena Sutra

Since the number ab = 10a+b, $(ab)^3 = (10a+b)^3 = 1000 a^3 + 3 (100) a^2 b + 3 (10) a b^2 + b^3$

The number $(ab)^3$ can be split into the following four parts

	1st	2nd	3rd	4th
$(a\,b)^3$	a^3	$a^2 b$	$a b^2$	b^3
		+	+	
		$2 a^2 b$	$2 a b^2$	
	a^3	$3 a^2 b$	$3 a b^2$	b^3

The above formula can be further simplified by substituting r = b/a, (r is called as common ratio) as shown below

	1st digit	2nd digit	3rd digits	4th digits
$(a\ b)^3$	a^3	a^3r	a^3r^2	a^3r^3
		+	+	
		$2a^3r$	$2a^3r^2$	
	a^3	$3a^3r$	$3a^3r^2$	a^3r^3

The working of this method to find the cube is as explained below

1. First part is obtained by taking cube of the first digit

2. Multiply the cube of first digit with r, r^2 and r^3 to get second, third and fourth parts respectively.

3. Double the second and the third number and put it down, under the second and third numbers in the second row. Finally add first and the second row

Example: Cube of 15

a =1 and b= 5

r = common ratio = b/a = 5/1 =5

1	5	25	125
	10	50	
1	15	75	125
3	**3**	**7**	**5**

Example: Cube of 17

a = 1 and b= 7

r = common ratio =7/1=7

1	7	49	343
	14	98	
1	21	147	343
4	**9**	**1**	**3**

Example: Cube of 32

a = 3 and b=2

r = common ratio = b/a = 2/3

27	27 x 2/3 =18	27 x 2/3 x2/3=12	27 x 2/3 x 2/3 x2/3 =8
	36	24	
27	54	36	8
32	**7**	**6**	**8**

Method 2 Nikhilam Sutra

The steps are as follows:

1. Take the deviation of the number to be cubed from its base.

2. Ist Term= (Number to be cubed + 2 x deviation from the base) x (sub base)^2

3. 2nd Term = { 3 x (deviation)^2 } x sub-base

4. (Deviation)^3

Example: Cube of 98

Ist Term	2nd Term	3rd Term
98+ 2 x (-2) = 94	3 x 4 =12	-8
94	**11**	**92**

Example: Cube of 26

Ist Term	2nd Term	3rd Term
(26+2 x 6) x 4	{ 3 x 36) x 2	216
152	216	216
175	**7**	**6**

Method: 3 Ekadhika Purvena

The steps for a two digit number are as follows:

1. 1st Term = Square of ten's digit x Ekadhika of ten's digit

2. 2nd Term = Square of ten's digit x Deviation , Deviation = 3 x unit digit -10

3. 3rd Term = 3 x ten's digit x square of unit digit

4. 4th Term = Cube of unit digit

Example: Cube of 26

Ist Term	2nd Term	3rd Term	4th Term
4 x 3	4 x (18-10)	3 x 2 x 36	6^3
12	32	216	216
17	5	7	6

Example: Cube of 93

Ist Term	2nd Term	3rd Term	4th Term
81 x 10	81 x -1	3 x 9 x 9	27
810	-81	243	27
800	100-81	243	27
804	3	5	7

Method: 4 Yavadunam Sutra

The steps are as follows:

1. Ist term = Number + (2 x excess or deficit from the base)

2. 2nd term = New excess or deficit x original excess or deficit

3. 3rd term = Cube of original excess or deficit

Example: Cube of 103

Ist Term	2nd Term	3rd Term
103 + 2 x 3	9 x 3	3^3
109	36	27

Find the cube of the following

19	107	121	21	28	31
44	87	207	315	54	104
29	34	63	46	63	98
99	13	99	102	105	17
71	83	96	95	98	95
14	15	33	55	99	56

Chapter -7 – Cube Root

The cube root of a number **"a"** whose cube is a^3 **is a**. We often have to deal with the cube roots of numbers while deriving side or radius of geometrical figures from their volume and other mathematical problems. Vedic Mathematics provides some excellent techniques to calculate the cube roots of numbers faster than the normal methods. In order to use the quick methods for finding the cube roots of the numbers you must remember cubes of the first ten natural number as described in the table below.

Cube Root

A	A^3
1	1
2	8
3	27
4	64
5	125
6	216
7	343
8	512
9	729

Following are the two techniques for finding the cube root of a number

1. Cube Root of a number having less than 7 digits

2. Cube root of a number having 7, 8 or 9 digits

Method 1: Cube Root of a number with less than 7 digits

Example1: Cube root 35937

5. Place a bar over the digits of the number from right to left, leaving 2 digits at a time

$$3 \quad \overline{5} \quad \overline{937}$$

6. The unit digit with bar is 7. Therefore the unit digit of the cube root is 3

7. The next bar falls on 5. The tens digit of the number is largest number whose cube root is less than or equal to 35 and as $3^3 < 3 < 4^3$

8. Hence the cube root is 33

Example 2: Cube Root of $1\ \overline{7}\ 2\ \overline{8}$

1. The unit digit with bar is 8. Therefore the unit digit of the cube root is 2

2. The next bar falls on 1. The tens digit of the number is largest number whose cube root is less than or equal to 1 and as $\qquad 1^3 < 1 < 2^3$

3. Cube root of 1728 is 12

Method 2: Cube root of a number having more than 7 digits but less than 10 digits

1. A number with 7, 8 or 9 digits will have digits in its cube root

2. Denote the left digit by L, the middle digit by M and the right digit by R

3. Subtract R^3 from the number and leave the last digit (0) obtained after subtraction

4. The middle digit is obtained by comparing $3 R^2$ M and the number obtained in step 3

5. The cube root of the number is L M R

Example1: Cube root of 3 3   0 7 6 1   6 1

1. The unit digit of the cube root is 1 and hence R =1.

2. Since $3^3 <33<4^3$ the left digit is 3

3. Subtract R^3 from the number and leave the last digit (0) obtained after subtraction

   33076161 -1 =33076160

   Leaving the last 0 the number obtained is 3307616

4. 3 M = 3307616, since 3M is contained in 3307616 both the number should end up in the same digit.

   The last digit in 3307616 is 6 and 3M should end with the same digit. 3M = 6

   M = 2

5. Since L=3, M =2 and R =1, the cube root is 321

Example2: Cube root of $\quad 1 \quad 4 \quad \overline{3} \quad 0 \quad 5 \quad \overline{5} \quad 6 \quad 6 \quad \overline{7}$

1. The unit digit of the number is 7, hence the unit digit of the cube root (R) is 3

2. As 64<143<216 , hence L = 5

3. Subtract R^3 from the number and leave the last digit (0) obtained after subtraction

$\quad$ 1 4 3 0 5 5 6 6 7

$\qquad\qquad\qquad$ 2 7

$\qquad$————————————

$\qquad$ 1 430 5 5 6 4

3 X 9 X M = 27 M is contained in 143055664 , so the both should end in same digit, hence M= 2.

4.   Since L=5, M=2 and R =3, the cube root is 523

Example3: Cube root of $4\ \overline{8}\ 2\ 2\ \overline{8}\ 5\ 4\ \overline{4}$

1.   R =4

2.   L= 3

3.   Subtract R^3 from the number and leave the last digit (0) obtained after subtraction

4 8 2 2 8 5 4 4

                    6   4

_____

  4 8 2 2 8   4   8   0

4.   48 M is contained in 4 8 2 2 8 4 8 , So M=1 or M=6. Now this is an ambiguous case and to avoid it divide the number by eight until you get an odd cube

8	4 8 2 2 8 5 4 4
8	6 0 2 8 5 6 8
	7 5 3 5 7 1

Now find the cube root of 7 5 $\overset{\cdot}{3}$ 5 $\overline{7}$ 1

5. R = 1

6. L = 9

7. Hence the cube root is 7 5 3 5 7 1 is 91

8. $\sqrt[3]{}$ 8 x 8 x 753571

    = 2 x 2 x 91 = 364

Example 4: Cube root of 1 9 0 6 6 2 4

Since this is an even number, hence to avoid the ambiguous case in extracting the cube root, we need to divide the number by 8 until we get an odd cube

8	1 9 0 6 6 2 4
8	2 3 8 3 2 8
	2 9 7 9 1

Grouping the number we get 2 $\overline{9\ 7}$ $\overline{9\ 1}$ and hence the cube root is 31

$\sqrt[3]{8 \times 8 \times 2\,9\,7\,9\,1} = 2 \times 2 \times 31 = 124$

Exercise: Find the cube root of the following

1. 636056

2. 8365427

3. 1061208

4. 8489664

5. 9800344

6. 33076161

7. 83453453

8. 143055667

9. 970299

10. 300763

Chapter -8

Finding out the decimal representation of fractions that have recurring decimal representation can be really time consuming if the denominator is large. This chapter shares some easy techniques from Vedic mathematics to deal with such fractions and these techniques are extremely fast.

Fractions and Recurring Decimals

Decimals Equivalents of Fractions

½ = .5
1/3 = .333...
1/4 = .25

1/3 = .333...
2/3 = .666...

¼ = .25
2/4 = ½ = .5
3/4 = .75

Fifths are very easy. Take the numerator (the number on top), double it, and stick a decimal in front of it.
1/5 = .2
2/5 = .4
3/5 = .6
4/5 = .8

There are only two new decimal equivalents to learn with the 6ths:
1/6 = .1666...
2/6 = 1/3 = .333...
3/6 = ½ = .5
4/6 = 2/3 = .666...
5/6 = .8333...

One-seventh is an interesting number:
1/7 = .142857142857142857...
For now, just think of one-seventh as: .142857, see if you notice any pattern in the 7ths:

1/7 = .142857...
2/7 = .285714...
3/7 = .428571...
4/7 = .571428...
5/7 = .714285...
6/7 = .857142...

Notice that the 6 digits in the 7ths ALWAYS stay in the same order and the starting digit is the only thing that changes!

If you know your multiples of 14 up to 6, it isn't difficult to, work out where to begin the decimal number. Look at this:
For 1/7, think "1 * 14", giving us .14 as the starting point.
For 2/7, think "2 * 14", giving us .28 as the starting point.
For 3/7, think "3 * 14", giving us .42 as the starting point.

For 4/14, 5/14 and 6/14, you'll have to adjust upward by 1:
For 4/7, think "(4 * 14) + 1", giving us .57 as the starting point.
For 5/7, think "(5 * 14) + 1", giving us .71 as the starting point.
For 6/7, think "(6 * 14) + 1", giving us .85 as the starting point.

8ths aren't that hard to learn, as they're just smaller steps than 4ths. If you have trouble with any of the 8ths, find the nearest 4th, and add .125 if needed:

1/8 = .125
2/8 = ¼ = .25
3/8 = .375
4/8 = ½ = .5
5/8 = .625
6/8 = ¾ = .75
7/8 = .875

9ths are almost too easy:
1/9 = .111...
2/9 = .222...
3/9 = .333...
4/9 = .444...
5/9 = .555...
6/9 = .666...
7/9 = .777...
8/9 = .888...

10ths are very easy, just put a decimal in front of the numerator:
1/10 = .1
2/10 = .2
3/10 = .3
4/10 = .4
5/10 = .5
6/10 = .6
7/10 = .7
8/10 = .8
9/10 = .9

11th are easy in a similar way, assuming you know your multiples of 9:

1/11 = .090909...
2/11 = .181818...
3/11 = .272727...
4/11 = .363636...
5/11 = .454545...

6/11 = .545454...
7/11 = .636363...
8/11 = .727272...
9/11 = .818181...
10/11 = .909090...

Recurring decimals

A fraction can be converted to a decimal by dividing the numerator by the denominator. The Vedic system gives us some easy and fast methods.

a) Denominator Ending in 9
'By one more than the one before' method is used where the denominator ends in **9**. For example in **1 / 19**, before the 9 in **1 / 19** is a **1** and one more than this is **2** and the word *'By'* tells us to divide by that **2**. So we keep dividing the numerator by this **2** (rather than dividing by 19).

1 / 19 = $0._1\dot{0}\ 5_1 2\ 63_1 1_1 5_1 7_1 89_1 47_1 3_1 6842\dot{1}$

We start with 0 and a decimal point. Then on dividing numerator 1 by 2 we get 0 remainder 1. The 0 is placed after the decimal point and the remainder digit is placed below and to the left of 0.

$0._1 0$

$0._1 05_1 2$

We now have 10 as the next number to be divided. So 10 divided by 2 = 5 and 5 divided by 2 = 2 rem 1. Now 2 is placed next to 5 and the remainder digit is placed below and to the left of the 2. making 12 as the next

$0._1 0\ 5_1 2\ 6\ 3_1 1$

$1 / 19 = 0._1\dot{0}\ 5_1 2$

$63_1 1_1 5_1 7_1 89_1 47_1 3_1 6842\dot{1}$

158

dividend.

12 divided by 2 = 6 and 6 divided by 2 = 3.

Continuing in this way the pattern begins to repeat after eighteen digits. At this stage the decimal is given its recurring dots to indicate that the pattern repeats.

Similarly, **11 / 19** = $0._1\dot{5}\,_17_189_147_13_168421_105_1263_1\dot{1}$

17 / 29 = $0._2\dot{5}\,_1862_20_26_28_19_16_155_217_124_11_23_27931_10_13_14_248_22_1\dot{7}$

A Short cut

We have worked out three recurring decimals: 1 /19, 11 / 19, 17 / 29 using *'By one more than the one before'* method. The first two have 18 recurring figures and the third has 28.
If we write out the three recurring decimals but with 1 / 19 and 11 / 19 in two rows of 9 figures and 17 / 29 with two rows of 14 figures we may notice something:

1 / 19 = $0._1\dot{0}\ 5_126\ 3_11_15_17$
$_18$

$\qquad 9_147_13_16\ 8\ 4\ 2\ \dot{1}$

17 / 29 = $0._2\dot{5}\ _18\ 6\ 2\ _20\ _26\ _28\ _19\ _16$
$_15\ 5\ _217\ _12$

$\qquad 4\ _11\ _23\ _27\ 9\ 3\ 1\ _10\ _13$

$\qquad _14\ _24\ 8\ _22\ _1\dot{7}$

11 / 19 = $0._1\dot{5}\ _17\ _18\ 9\ _14\ 7\ _13$
$_16\ 8$

$$4\ 2\ 1\ _10\ 5_12\ 6\ 3_1\dot{1}$$

In each case the total of every column is the same: every column adds up to 9. This means that when we have got half way through and can write down the second half from the first half. We just take every figure in the first half from 9 and this gives the second half.

To find out the half way (**H**), we subtract the numerator from the denominator: $19 - 1 = 18$. When 18 comes up then we are half way through. In recurring decimal for $1 / 19$ we see 18 at the end of the first line.

Similarly for $11 / 19$, the H = 19-11 = 8 and for $17 / 29$, H = 29-17 = 12

If the difference of the numerator and denominator comes up in the recurring decimal then we are half way through and we can get the second half by taking all the figures in the first half from 9. (The difference of numerator and denominator does not always appear however.)

Convert **9 / 39** to decimal

Using *'By one more than the one before'* the divisor fir this is 3 + 1 = 4and H = 39-9 = 30. Now dividing 9 by 4 we get $9 / 39 = 0._12 3_10$
The first three steps give us 12, 3 and 30 and so we find that the half way number has come up after just 3 figures. So we just take each of the first three figures from 9 to get the complete answer:

9 / 39 = $0._1\dot{2}\ 3_30 7 6\ \dot{9}$

160

Convert **10 / 39** to decimal

Using *'By one more than the one before'* the divisor fir this is $3 + 1$

$= 4$ and $H = 39-10 = 29$. Now dividing 10 by 4 we get $10 / 39 = 0._2\dot{2}$

$_25_1641_1\dot{0}$ but in this case the half way number doesn't come up.

10 / 39 $= 0._2\dot{2}\ _25_1641_1\dot{0}$

Exercise

Find the recurring decimal for each of the following
25 / 29   24 / 39   29 / 39   3 / 49   44 / 69   44 / 79   1 / 99   1 / 9

Proportionately

The special method we have been using can only be applied when the denominator ends in 9. We apply *Proportionately* formula to make this method work for other denominators. If the denominator does not end in 9 then we can multiply the top and bottom of a fraction by a number which can convert the denominator to a number that ends in 9 without changing its value.

Convert **7 / 13** to a recurring decimal

Here denominator ends in 3 and doesn't end in 9. However we know we can multiply the top and bottom of this fraction by 3 to get 9 in the denominator.

So, $\dfrac{7x3}{13x3}$ = 21 / 39.

Now we can find the decimal for 21 / 39 as before using *'By one more than the one before'* the divisor for this is 3 + 1 = 4 and H = 39-21 =18

21 / 39 = $0._1\dot{5}\,_33_18461$

The half way number is 18 and it comes up after 3 figures.

Convert **1 / 7** to a recurring decimal

$\dfrac{1x7}{7x7}$ = 7 / 49, *By one more than the one before'* the divisor for this is 4 + 1 = 5 and H = 49-7 =42

7 / 49 = $0._21_14_4285\,\dot{7}$

162

The half way number is 42 and it comes up after 3 figures.

Exercise

Find the recurring decimal for each of the following
1 / 13   2 / 13   5 / 23   17 / 33   9 / 11   3 / 17
3 / 7
Find the correct 4 decimal places
18 / 59   67 / 89   100 /109 $1\dfrac{3}{7}$ 20 / 13   99 / 49

Longer Numerator

This method is for any numerator and not just a whole number between 0 and the denominator.

Find $1.23 \div 19 = 0._106_147_13.....$

We find 1 / 19 and simply add 2 when we deal with the tenths and add 3 when we deal with the hundredths.

$$1.23 / 19 = 0._106_147_13.....$$

We begin with 2 dividing 1

$$0._10$$

$$0._106$$

Next instead of dividing 10 by 2 we divide 12 by 2 because we add on the 2 of 1.23. This gives 6 which we put down.

$$0._106_14$$

$$0._106_147_{....}$$

Then the 3 of 1.23 is added to 6 , so

we divide 9 by 2 and put down $_14$

From here on we just divide by 2 as
before and 14 divided by 2 = 7

Find 2345 $\div$ 49 = $_34_37._28_35....$

23 divided by 5 = 4 rem 3 , put down $_34$ $_34$

Now add 4 of 2345 to 34 to get 38, 38 divided by 5 $_34_37$
= 7 rem 3, put down $_37$.

                                $_34_37._28$

Now add 5 of 2345 to 37 to get 42, 42 divided by 5
= 8 rem 2 and place $._28$. $_34_37._28_35...$

From here we divide as normal, 28 by 5 = 5 rem 3
and place $_35$

Exercise

Divide the following

$5.67 \div 19$	$67.8 \div 29$	$555 \div 39$	$0.0135 \div 79$	$321 \div 13$
$33 \div 9.5$	$19.19 \div 59$	$18.88 \div 19$		

Denominators Ending in 8, 7, 6 2

If this scenario we can use the same method as for denominator ending in 9 but multiply the last figure at each step by 2,3,4...8 before dividing

$5/28 = ._2 1 \,_1 78 \,_1 5\, 7\, 4$

In this case AF is .5/3 so we will start by dividing 3 into .5. This gives 1 remainder 2. As mentioned above we will double the last number as 28 is two below 30 and add it to the remainder. Hence the new number to be divided by 3 becomes 22. This gives 7 remainder 1. Now we will keep following the same steps to get numbers at subsequent decimal places

---

Divide the following

$5.67 \div 28$ $67.8 \div 37$ $555 \div 58$ $0.0135 \div 77$ $321 \div 87$ $33 \div 97$ $19.19 \div 67$ $18.88 \div 37$

Denominators Ending in 1

Suppose we have to find out 3/31

3 in the numerator can be written as 2 +.9999999..

So divide 3 into 2 then subtract quotient digit from 9 and repeat the same steps.

3/31 = 0.0967741935483871

5/21 = 0.2380952380952381

Divide the following

5.67 ÷ 21   67.8÷31   555 ÷51   0.0135 ÷ 71   321 ÷ 81   33 ÷ 91   19.19 ÷ 61   18.88 ÷ 31

Denominators ending in 2, 3 and 4

3/42 = 0.0714285714285714

4/43 = 0.0930232558139535

5/44= 0.1136363636363636

The procedure followed is same as that for denominator ending in 1, the only difference is that we have to multiply the last digit (Quotient digit) by 2,3,4 for denominator ending with 2,3 and 4 before taking the complement from 9.

166

Divide the following

5.67 ÷ 22   67.8 ÷ 34   555 ÷ 52   0.0135 ÷ 73   321 ÷ 81

33 ÷ 91   19.19 ÷ 61   18.88 ÷ 31

WORKING 2, 3 ETC. FIGURES AT A TIME

$2/199 = 0.\ 01\ _100\ 50\ 25\ _112\ 56.......$

The Auxiliary fraction is 2/2, with 2 into 2 goes 01 and remainder 0 so we put down .01.

2 into 01 goes 00, which we put down $_100$

This gives 100 and 2 into 100 goes 50

Then 2 into 50 is 25 and son on so forth

$108/2001 = .\,_1053\ 973\ 013\ 493$

AF = .107/2

$107 ÷ 2 = 53$ remainder 1 so we put down $0.\,_1053$

Subtract each of the figures from 9 to get 946, 2 goes into 1946 = 973 and so on so forth

Divide the following

48/199   37/299   535/4999   70/233   57/201

222/19999 63/298

HCF and LCM

Highest Common Factor (HCF)

The highest common factor of two or more numbers is the highest number which can divide those numbers exactly.

For Example, the highest common factor of 8 and 4 is 4. The highest of the factors that are common between 8 and 4 is 4, hence 4 is the HCF. Similarly the HCF of 10 and 15 is 5.

If the HCF of two numbers is 1 then they are called as co-prime

Prime Factors to find HCF

By expressing the numbers in terms of prime factors the HCF can be found by taking all those factors which are common to both

 ➢ **HCF of two co-primes is 1**

Example: HCF of 420 and 36

2	420
2	210
3	105
5	35
	7

2	36
2	18
2	9
3	3

Common Factors are 2x2x3 =12

Therefore 12 is the HCF

Exercise

Find the HCF by prime factors

210,750	303,309	528,888	91,280	189, 882, 1071
756,1764,2268		108,162,270	784, 1232, 1904	

Elimination and Retention method to find HCF

Example: Find the HCF of 168 and 432

168 and 432 are divisible by 2, elimination leaves 84 and 216. Both of these are divisible by 4, leaving 21 and 54. Dividing by 3, we are left with 7 and 18. This last pair of number is co prime and so the HCF is 2x4x3 =24

```
2 |168   432
  |_____
4 | 84   216
  |_____
3 | 21   54
  |_____
  | 7    18
```

Exercise

Find the HCF by elimination and retention

175,245	1428, 2856	704, 320	112, 84	630, 360
868, 372				

LCM (Lowest Common Multiple)

The lowest common multiple of two or more numbers is the lowest number into which those numbers can divide

Vertically and crosswise method for finding LCM

Finding the lowest common multiples of two numbers is almost the same as finding the highest common factor but includes the two prime factors in the product. The following example illustrate the method

> **LCM of two co-primes is always equal to their products**

> **If a and b are two numbers then, a x b = HCF of a and b x LCM of and b**

Example: Find LCM of 12 and 30

Extract common factors until the two numbers remaining are co prime. Since 6 is a factor of both 12 and 30, we divide 6 into 12 and 30, leaving 2 and 5. These are co prime. The LCM is found by cross multiplying 12x5 or 30x2=60

```
6│12  30
 └──────
   2   5
```

LCM = 30 X2 =60

Example: Find LCM of 96 and 72

Here 8 is a common factor and upon dividing 96 and 72, leaves 12 and 9. These can further be divided by 3, leaving 4 and 3 which are co prime. The LCM is then cross-product 96 x

3 or 72 x 4=288

```
8│ 96  72
 └────────
3│ 12  9
 └────────
   4   3
```

Exercise

Find the LCM

308,309   210, 750   303, 309   528, 888   68, 92   91, 280

Exercise:

1. What is the least length of rope which can be cut into pieces which are 1ft 6 in long and into pieces which are 1ft 3 in long?

2. Either by walking with strides of 96cms or with strides of 90cm, I take an exact number of steps to cross a road. What is the least possible width of the road?

3. Three chimes on a clock strike at intervals of 0.4, 0.6 and 0.9 seconds respectively. If they start together , how long will it be before they strike together again

4. Alice, Amy and Anna run up a flight of stairs, all starting from the bottom, Alice take two steps at a time, Amy runs three and Anna takes five. After how many stairs will they all land on the same one?

5. Find a pair of number between 100 and 130 which have 14 as their HCF

6. A rectangular block measuring 6cm by 12 cm by 15cm is to be cut up into cubes. Find the smallest possible number of cubes

7. The length, breadth and height of a room are 1050 cm, 750 cm, and 425 cm respectively. Find the length of the longest tape which can measure the three dimensions of the room exactly

8. Determine the longest tape which can be used to measure exactly the lengths 7 meters, 3 meters 85 cm and 12 meters 95 cm.

9. The circumferences of four wheels are 50cm, 60cm, 75cm and 100cm. They start moving simultaneously. What least distance should they cover so that each wheel makes a complete number of revolution

10. The HCF of two numbers is 23 and their LCM is 1449. If one of the number is 161 find the other

11. Can two numbers have 16 as their HCF and 204 as their LCM? Give reason

12. Find the greatest number which divides 285 and 1249, leaving remainders 9 and 7 respectively.

13. Reduce 289/391 to the lowest terms

14. The traffic lights at three different road crossings change after every 48 seconds, 72 seconds and 108 seconds. If they start changing simultaneously at 8.am. , after how much time will they change again simultaneously

An electronic device makes a beep after every 15 minutes. Another device makes a beep after every 20 minutes. They beeped together at 6 am. At what time will they next beep together?

Ratio and Proportion

The ratio of two quantities of the similar nature and in the same units is a fraction that shows how many times the one quantity is of the other

The ratio of two quantities p and q (q≠0) is p ÷ q or, p/q and is denoted by p: q, p and q are called the terms of the ratio; p is called the first term or antecedent and q is known as second term or consequent.

A ratio p: q is in simplest form if p and q have no common factors

Example: The angles of a triangle are in the ratio of 2:3:4. Determine the angles

Solution: Let the angles be 2x, 3x and 4x.

Since the sum of the angles of the triangle is 180 degrees.

$2x+3x+4x = 180$

$9x = 180, x = 20$

Hence the angles are 40, 60 and 80 degrees

Example: The ratio between two numbers is 3:4. If their LCM is 180, find the numbers.

Solution: Let the numbers be 3x and 4x. LCM of 3x and 4x is 12x.

12 x =180

Hence x= 180/12=15

Problems:

1. The two numbers are in the ratio 7:11. If 7 is added to each of the numbers, the ratio becomes 2:3. Determine the numbers (Answer-49,77)

2. In a class, one out of every six student fails. If there are 42 students in the class, how many pass? (Answer-35)

Direct proportion implies if on quantity is increased the other quantity is increased in the same ratio. E.g. if three books cost Rs 30 then we would expect 9 books to cost Rs 90. As the number of books increases the cost increases in the same ration

Problems:

1. If 6 Litres of liquid have mass of 30 Kg, find the mass of 16 litres of the liquid.(Answer-80 Kg)
2. If 5 planes can land at Heathrow airport every 6 minutes, find how many planes can land in half an hour. (Answer-25)
3. If 12 electric bulbs cost Rs 7.46. How many can be bought for Rs 9.45 .(Answer-15)

4. The population of Great Britain is about 56 Million and the area is about 100,000 square miles. If the USA, with an area of about 3,700,000 square miles had the same number of people per square mile, what would be its population?
(Answer-11,340,000)

Inverse variation implies if one quantity increases in the same ratio as the other quantity decreases. If four men take 10 days to paint the wall clearly the time taken by 10 men would be less.

Example: If 15 men can build a brick wall in 14 hours, how long will it take with 21 Men?

Solution: Let x be the number of hours taken by 21 men

15 /14 = x/21

x = 15 x14/21

x= 10

Problems:

1. A 3 ton lorry removes a mound of earth in 15 journeys. How many journeys would a 5 ton lorry have to make (Answer-9)?

2. A certain ditch can be dug by 8 men in 3 days. How long will it take with 6 men (Answer-4 days)?

3. If the sun can dry 12 shirts on a washing line in one and a half hour, how long will it take to dry fifteen shirts (Answer- 1.30 hrs)?

4. A section of wall can be covered with 72 tiles each 4 in by 4 in. How many tiles are needed if the tiles used are 6 in by 6 in (Answer – 32 tiles)

5. 30 men can mend a section of railway track in 24 days. How long will it take 18 men to mend it (Answer- 40 days

6. The winning relay team in a high school sports competition clocked 48 min for a distance of 13.2km. Its runners A, B, C and D maintained speeds of 15 km/hr, 16km/hr, 17km/hr and 18 km/hr respectively. What is the ratio of time of time taken by B to the time taken by D? (Answer -8/9)

7. A and B walks from X to Y, a distance of 27 km at 5 km/h and 7 km/h respectively. B reaches Y and immediately turns meeting A at Z. What is the distance from X to Z? (Answer – 22.5 km)

8. Ram went from Delhi to Shimla via Chandigarh by car. The distance from Delhi to Chandigarh is ¾ times the distance from Chandigarh to Shimla. The average speed from Delhi to Chandigarh was half as much as that from Chandigarh to Shimla. If the average speed for the entire journey was 49 km/h, what was the average speed from Chandigarh to shimla? (Answer-42 km/hr)

9. A man can do as much work in one day as a woman can do in 2 days. A child does one-third the work in a day as a woman. If an estate owner hires 39 pairs of hands – men, women and children in the ratio 6:5:2 and pays them in all Rs 1,113 at the end of the day's work, what must the daily wages of a child be, if the wages are proportional to the amount of work done (Answer –Rs 7)

10. Two towns A and B are 100 km apart. A school is to be built for 100 students of town B and students of town A. Expenditure on transport is Rs 1.20 per kilometer per student If the total expenditure on transport by all 130 students is to be as small as possible, then the school should be built at :

      a.   3 3 Km from town A

      b.   Town B

      c.   33 km from town B

      d.   Town A

Chapter - 9

Quantitative Aptitude

Try problems on quantitative aptitude in the section below and find out (we have illustrated in some cases) how they can be solved quickly using Vedic mathematics.

Carol and John leave the same camp and run in opposite directions. Carol runs 99 mile per second (mps) and John runs 98 mps. How far apart are they in miles after 1 minute 38 seconds?

S1 = Distance (Carol) = speed (carol) x time= 99 x 98 = 9702
Vedic technique will help

S2 = Distance (John) = speed (John) x time = 98 x98= 9604
(Multiplication of numbers close to the base)

Total distance (s) between Carol and John = 9702 + 9604 = 19307

A person starts from hill top A, goes downhill, then along a plain and finally climbs to hill top B. He takes a total of 4 hours for the journey travelling downhill @72kmph, on plains @63kmph and uphill @56kmph. However while returning back, from B to A, he takes 4 hours 40 mins with speeds being same. Find distance AB

Let x be the distance from A to planes, and y the distance travelled along the plain and z be the distance from plain to uphill(to point B).

$X/72 + Y/63 + Z/56 = 4$, 7X+8Y+9Z=72 X7 X4

$Z/72+Y/63+X/56=4.4$, 7Z+8Y+9X=72 X7 X4.4

X=63 X 4.2 } $\longrightarrow$ Vedic technique (Vertically crosswise will help to do multiplication faster)

X+Y+Z=264.4

A wall is built by 17 men in 24 days. In how many days can 19 men do the work if hours per day are reduced in the ratio 5:4?

18 men can do the work in 17 x24/19 x 5/4= 510/19= 26.842 days} $\longrightarrow$ Vedic technique for recurring decimals used for denominator ending in nine will help in this scenario

180

The diameter of the driving wheel of a bus in 111 cm. How many revolutions per minute must the wheel make in order to keep a speed of 66 kmph?

Distance to be covered in 1 min. = [66*1000/60]m = 1100m

Circumference of the wheel = [2*22/7* 111/200]m = 3.48 m

Use Vedic technique for 44 x 111 =4884

Number of revolutions per min. = 1100/3.48 = 316.09

Eleven straight lines are drawn in a plane such that no two of them are parallel and no three of them are concurrent. An ellipse is now drawn in the same plane such that all the points of intersection of all the lines lie inside the ellipse. What is the number of non-overlapping regions into which the ellipse is divided?

If n lines are drawn in the given manner, the required number of regions will be

$$= [n \times (n+1)]/2 + 1$$
$$= (11 \times 12) / 2 + 1$$
$$= 67$$

Converting Kilos to pounds

- ***86 kilos into pounds:***
  1 Kg = 2.2 pounds

Step one, multiply the kilos by TWO.

To do this, just double the kilos.

86 x 2 = 172

Step two, divide the answer by ten.
To do this, just put a decimal point one place in from the right.

172 / 10 = 17.2

Step three, add step two's answer to step one's answer.

172 + 17.2 = 189.2

86 Kilos = 189.2 pounds

- ***50 Kilos to pounds:***

Step one, multiply the kilos by TWO.
To do this, just double the kilos.

50 x 2 = 100

Step two, divide the answer by ten.
To do this, just put a decimal point one place in from the right.

100/10 = 10

Step three, add step two's answer to step one's answer.

100 + 10 = 110

50 Kilos = 110 pounds

<u>Adding Time</u>

Simple way to add hours and minutes together: Let's add 1 hr and 35 minutes and 3 hr 55 minutes together.

Make the 1 hr 35 minutes into one number, which will give us 135 and do the same for the other number, 3 hours 55 minutes, giving us 355.

Now you want to add these two numbers together:

135 + 355 = 490, So we now have a sub total of 490.

What you need to do to this and all sub totals is **add the time constant of 40.**

Note – The reason for adding 40 is – The decimal number system follows the cycle of 100 where as the time follows cycle of 60.

No matter what the hours and minutes are, just add the 40 time constant to the sub total.

490 + 40 = 530, So we can now see our answer is 5 hrs and 30 minutes!

<u>Temperature Conversions</u>

Fahrenheit to Celsius and vice versa.

The answer you will get will not be an exact one, but it will give you an idea of the temperature you are looking at.

Fahrenheit to Celsius: c= 5/9 (F-32)

Take 30 away from the Fahrenheit, and then divide the answer by two.
This is your answer in Celsius.

Example:
74 Fahrenheit - 30 = 44. Then divide by two, 22 Celsius. So, 74 Fahrenheit = 22 Celsius.

Celsius to Fahrenheit just do the reverse:

Double it, and then add 30.

30 Celsius double it, is 60, then add 30 is 90. So, 30 Celsius = 90 Fahrenheit

Remember, the answer is not exact but it gives you a rough idea.

<u>**Converting Kilometers to Miles**</u>

The formula to convert kilometers to miles is(number of kilometers / 8) X 5

- **80 kilometres into miles**

80/8 = 10, multiplied by 5 is 50 miles!

- **40 kilometres**

40 / 8 = 5, 5 X 5= 25 miles

What approximate value should come in place of the question mark (?) in the following questions ?

$(4863 + 1174 + 2829) \div 756 = ?$

(A) 18

(B) 16

(C) 12

(D) 9

(E) 22

Ans : (C)

$37 \cdot 35 + 13 \cdot 064 \times 3 \cdot 46 = ?$

(A) 89

(B) 83

(C) 76

(D) 79

(E) 85

Ans : (B)

$54 \times 746 \div 32 = ?$

(A) 1259

(B) 1268

(C) 1196

(D) 1248

(E) 1236

Ans : (A)

The ratio of ducks and frogs in a pond is 37 : 39 respectively. The average number of ducks and frogs in the pond is 152. What is the number of frogs in the pond ?

(A) 148

(B) 152

(C) 156

(D) 144

(E) None of these

Ans : (C)

In how many different ways can the letters of the word 'ARISE' be arranged ?

(A) 90

(B) 60

(C) 180

(D) 120

(E) None of these

Ans : (D)

A milkman sells 120 litres of milk for Rs. 3360 and he sells 240 litres of milk for Rs. 6120. How much concession does the trader give per litre of milk, when he sells 240 litres of milk ?

(A) Rs. 2

(B) Rs. 3·5

(C) Rs. 2·5

(D) Rs. 1·5

(E) None of these

Ans : (C)

When 3626 is divided by the square of a number and the answer so obtained is multiplied by 32, the final answer obtained is 2368. What is the number?

(A) 7

(B) 36

(C) 49

(D) 6

(E) None of these

Ans : (A)

The sum of the two digits of a two digit number is 14. The difference between the first digit and the second digit of the two digit number is 2. What is the product of the two digits of the two digit number ?
(A) 56
(B) 48
(C) 45
(D) Cannot be determined
(E) None of these
Ans : (B)

A car runs at the speed of 50 kms per hour when not serviced and runs at 60 kms/hr. when serviced. After servicing the car covers a certain distance in 6 hours. How much time will the car take to cover the same distance when not serviced ?
(A) 8·2 hours
(B) 6·5 hours
(C) 8 hours
(D) 7·2 hours
(E) None of these
Ans : (D)

Practice Questions

1. What is the remainder when 4^{96} is divided by 6?
   a) 0 b) 2 c) 3 d) 4

2. The reminder when $(15^{23}+23^{23})$ is divided by 19?
   a) 4 b) 15   c) 0 d 18

Hint – a^n+b^n is divisible by a+b when n is odd.

3. Martha is 48 years old and her husband hug is 56. Their daughter is twice the highest common factor of her parent's age. How old is she? (Answer – 16 Yrs)

4. Ram and Shyam meet at the library on a Saturday. After this, Ram returns every three days and Shyam returns every 5 days. After how many days will they meet again at the library? Which day of the week is this (Answer- 15, Sunday)

5. A boy spent ½ of his money in one shop and $1/4^{th}$ of it in another. He then had 40 Rs left. How much did he begin with (Answer- 1.60 Rs)

6. When fully load a lorry can carry 3 ¼ tons of sand. Find how much it can deliver in carrying 9 ½ loads (Answer – 30 7/8 tons)

7. My father is 180 cm tall. I am 4 cm taller than 2/3 of his height .How tall am I (Answer – 124 cms)

8. A man wishes to build an extension to his house and council planning regulations allow an increase of 1/5 original floor space, if the floor space in his house is 1600 sq feet. What is the largest floor space allowed for the extension (Answer -320 sq feet).

9. $7^{6n}-6^{6n}$, where n is an integer >0 , is divisible by

a) 13   b) 127   c) 559   d) None of these

   Hint - $(7^{3n} - 6^{3n}) (7^{3n} + 6^{3n})$ hence the answer is 127

10. Anita had to do a multiplication. Instead of taking 35 as one of the Multiplier, she took 53. As a result, the product went up by 540.What is the new product (Answer -1590)

11. If one kg of pure milk contains 0.45 kg of fat. How much fat is there in 14.4 kg of milk (Answer-3.528 Kgs)

12. A red light flashes 3 times per minute and a green light flashes 5 times in two minutes at regular intervals. If both lights start flashing at the same time, how many times do they flash together in each hour? (Answer-30)

13. Find the area of a square part whose perimeter is 384 m (Answer -9216)

14. One side of the square field is 179 meter. Find the cost of raising a lawn on the field at the rate of Rs 1.50 per square meter. (Answer – Rs 48061.50)

15. Three pieces of cake of weight 4 ½ lbs , 6 ¾ lbs and 7 1/5 lbs respectively are to be divided into parts of equal weights, Further each part must be as heavy as possible. If one such part is served to each guest, then what is the maximum number of guests that could be entertained? (Answer -41)

16. An error 2% in excess is made while measuring the side of a square. The percentage of error in the calculated area of the square is: (Answer - 4.04%)

17. A man walked diagonally across a square lot. Approximately, what was the percent saved by not walking along the edges? (Answer-30)

18. A rectangular field is to be fenced on three sides leaving a side of 20 feet uncovered. If the area of the field is 680 sq. feet, how many feet of fencing will be required? (Answer - D)

19. A two-digit number is such that the product of the digits is 8. When 18 is added to the number, then the digits are reversed. The number is: (Answer- 24)

20. The difference between a two-digit number and the number obtained by interchanging the positions of its digits is 36. What is the difference between the two digits of that number? (Answer- 4)

21. The least perfect square, which is divisible by each of 21, 36 and 66 is. (Answer – 213444)

22. 1.5625 = ? (Answer – 1.25)

23. A group of students decided to collect as many paise from each member of group as is the number of members. If the total collection amounts to Rs. 59.29, the number of the member is the group is. (Answer 77)

24. From a group of 7 men and 6 women, five persons are to be selected to form a committee so that at least 3 men are there on the committee. In how many ways can it be done? (Answer – 756)

25. In a shower, 5 cm of rain falls. The volume of water that falls on 1.5 hectares (10,000 sq m) of ground is: (Answer – 750 m^3)

26. 66 cubic centimeters of silver is drawn into a wire 1 mm in diameter. The length of the wire in meters will be: (Answer – 84 meters)

27. A hollow iron pipe is 21 cm long and its external diameter is 8 cm. If the thickness of the pipe is 1 cm and iron weighs 8 g/cm3, then the weight of the pipe is. (Answer -3.696 Kg)

28. A boat having a length 3 m and breadth 2 m is floating on a lake. The boat sinks by 1 cm when a man gets on it. The mass of the man is (Answer - 60 Kg)

29. A cistern 6m long and 4 m wide contains water up to a depth of 1 m 25 cm. The total area of the wet surface is. (Answer - 49 m^2)

30. Running at the same constant rate, 6 identical machines can produce a total of 270 bottles per minute. At this rate, how many bottles could 10 such machines produce in 4 minutes? (Answer - 1800)

31. If a quarter kg of potato costs 60 paise, how many paise will 200 gm cost? (Answer – 48 Paise)

32. If 5a = 3125, then the value of 5(a - 3) is: (Answer – 25)

33. If m and n are whole numbers such that $m^n = 121$, the value of $(m - 1)n + 1$ is: (Answer – 1000)

34. A two-digit number is such that the product of the digits is 8. When 18 is added to the number, then the digits are reversed. The number is: (Answer - 24)

35. The reflex angle between the hands of a clock at 10.25 is (Answer – $197\frac{1}{2}^0$)

36. A watch which gains 5 seconds in 3 minutes was set right at 7 a.m. In the afternoon of the same day, when the watch indicated quarter past 4 o'clock, the true time is.
(Answer - 4 pm)

37. How much does a watch lose per day, if its hands coincide every 64 minutes? (Answer - 32 8/11 min)

38. At what time between 5.30 and 6 will the hands of a clock be at right angles? (Answer – 43 7/11 past 5)

39. A vessel is filled with liquid, 3 parts of which are water and 5 parts syrup. How much of the mixture must be drawn off and replaced with water so that the mixture may be half water and half syrup? (Answer – 1/5)

40. A can contains a mixture of two liquids A and B is the ratio 7 : 5. When 9 litres of mixture are drawn off and the can is filled with B, the ratio of A and B becomes 7 : 9. How many litres of liquid A was contained by the can initially? (Answer 21 litres of A)

41. A milk vendor has 2 cans of milk. The first contains 25% water and the rest milk. The second contains 50% water. How much milk should he mix from each of the containers so as to get 12 litres of milk such that the ratio of water to milk is 3 : 5? (6 litres, 6 litres)

42. In what ratio must a grocer mix two varieties of pulses costing Rs. 15 and Rs. 20 per kg respectively so as to get a mixture worth Rs. 16.50 kg? (Answer – 7:3)

43. A dishonest milkman professes to sell his milk at cost price but he mixes it with water and thereby gains 25%. The percentage of water in the mixture is. (Answer – 20%)

44. How many kilogram of sugar costing Rs. 9 per kg must be mixed with 27 kg of sugar costing Rs. 7 per kg so that there may be a gain of 10% by selling the mixture at Rs. 9.24 per kg? (Answer – 63 Kg)

45. A container contains 40 litres of milk. From this container 4 litres of milk was taken out and replaced by water. This process was repeated further two times. How much milk is now contained by the container? (Answer – 29.16 litres)

46. In what ratio must water be mixed with milk to gain $16\frac{2}{3}$% on selling the mixture at cost price? (Answer 1:6)

47. In what ratio must a grocer mix two varieties of tea worth Rs. 60 a kg and Rs. 65 a kg so that by selling the mixture at Rs. 68.20 a kg he may gain 10%? (Answer 3:2)

48. 8 litres are drawn from a cask full of wine and is then filled with water. This operation is performed three more times. The ratio of the quantity of wine now left in cask to that of water is 16 : 81. How much wine did the cask hold originally? (Answer – 24 litres)

49. A merchant has 1000 kg of sugar, part of which he sells at 8% profit and the rest at 18% profit. He gains 14% on the whole. The quantity sold at 18% profit is: (Answer -600 Kg)

50. A grocer has a sale of Rs. 6435, Rs. 6927, Rs. 6855, Rs. 7230 and Rs. 6562 for 5 consecutive months. How much sale must he have in the sixth month so that he gets an average sale of Rs. 6500? (Answer – Rs. 4991)

51. The average monthly income of P and Q is Rs. 5050. The average monthly income of Q and R is Rs. 6250 and the average monthly income of P and

R is Rs. 5200. The monthly income of P is: (Answer – Rs 4000)

52. A car owner buys petrol at Rs.7.50, Rs. 8 and Rs. 8.50 per litre for three successive years. What approximately is the average cost per litre of petrol if he spends Rs. 4000 each year? (Answer Rs 7.98)

53. In Arun's opinion, his weight is greater than 65 kg but less than 72 kg. His brother does not agree with Arun and he thinks that Arun's weight is greater than 60 kg but less than 70 kg. His mother's view is that his weight cannot be greater than 68 kg. If all are them are correct in their estimation, what is the average of different probable weights of Arun? (Answer -67Kg)

54. A library has an average of 510 visitors on Sundays and 240 on other days. The average number of visitors per day in a month of 30 days beginning with a Sunday is: (Answer – 285)

55. A pupil's marks were wrongly entered as 83 instead of 63. Due to that the average marks for the class got increased by half (1/2). The number of pupils in the class is: (Answer- 40)

56. A, B and C jointly thought of engaging themselves in a business venture. It was agreed that A would invest Rs. 6500 for 6 months, B, Rs. 8400 for 5 months and C, Rs. 10,000 for 3 months. A wants to be the working member for which, he was to receive 5% of the profits. The profit earned was Rs. 7400. Calculate the share of B in the profit. (Answer – Rs 2660)

57. A and B entered into partnership with capitals in the ratio 4 : 5. After 3 months, A withdrew of his capital and B withdrew $\frac{1}{5}$ of his capital. The gain at the end of 10 months was Rs. 760. A's share in this profit is: (Answer – Rs 330)

58. A, B, C rent a pasture. A puts 10 oxen for 7 months, B puts 12 oxen for 5 months and C puts 15 oxen for 3 months for grazing. If the rent of the pasture is Rs. 175, how much must C pay as his share of rent? (Answer – Rs 45)

59. A and B started a business in partnership investing Rs. 20,000 and Rs. 15,000 respectively. After six months, C joined them with Rs. 20,000. What will be B's share in total profit of Rs. 25,000 earned at the end of 2 years from the starting of the business? (Answer – Rs 7500)

60. Two students appeared at an examination. One of them secured 9 marks more than the other and his marks was 56% of the sum of their marks. The marks obtained by them are: (Answer 42 and 33)

61. A fruit seller had some apples. He sells 40% apples and still has 420 apples. Originally, he had: (Answer – 700 apples)

62. In a certain school, 20% of students are below 8 years of age. The number of students above 8 years of age is $\frac{2}{3}$ of the number of students of 8 years of age which is 48. What is the total number of students in the school? (Answer – 100)

63. A student multiplied a number by 3/5 instead of 5/3 What is the percentage error in the calculation? (Answer 64%)

64. In an election between two candidates, one got 55% of the total valid votes, 20% of the votes were invalid. If the total number of votes was 7500, the number of valid votes that the other candidate got, was: (Answer – 2700)

65. A pupil's marks were wrongly entered as 83 instead of 63. Due to that the average marks for the class got increased by half (1/2). The number of pupils in the class is: (Answer - 40)

67. Two tailors X and Y are paid a total of Rs. 550 per week by their employer. If X is paid 120 percent of the sum paid to Y, how much is Y paid per week? (Answer – 250)

68. Rajeev buys good worth Rs. 6650. He gets a rebate of 6% on it. After getting the rebate, he pays sales tax @ 10%. Find the amount he will have to pay for the goods. (Answer - Rs. 6876.10)

69. The population of a town increased from 1,75,000 to 2,62,500 in a decade. The average percent increase of population per year is: (Answer – 5%)

70. A father said to his son, "I was as old as you are at the present at the time of your birth". If the father's age is 38 years now, the son's age five years back was: (Answer – 14 Years)

71. Six years ago, the ratio of the ages of Kunal and Sagar was 6 : 5. Four years hence, the ratio of their ages will be 11 : 10. What is Sagar's age at present? (Answer – 16 Years)

72. At present, the ratio between the ages of Arun and Deepak is 4 : 3. After 6 years, Arun's age will be 26 years. What is the age of Deepak at present ? (Answer – 15 Years)

73. Ayesha's father was 38 years of age when she was born while her mother was 36 years old when her brother four years younger to her was born. What is the difference between the ages of her parents? (Answer – 6 Years)

74. A boat running upstream takes 8 hours 48 minutes to cover a certain distance, while it takes 4 hours to cover the same distance running downstream. What is the ratio between the speed of the boat and speed of the water current respectively? (Answer – 8:3)

75. In one hour, a boat goes 11 km/hr along the stream and 5 km/hr against the stream. The speed of the boat in still water (in km/hr) is: (Answer -8 km/hr)

76. A man can row at 5 kmph in still water. If the velocity of current is 1 kmph and it takes him 1 hour to row to a place and come back, how far is the place? (Answer- 2.4 km)

77. A boat covers a certain distance downstream in 1 hour, while it comes back in $1\frac{1}{2}$ hours. If the speed of the stream be 3 kmph, what is the speed of the boat in still water? (Answer 15 km/hr)

78. A man can row three-quarters of a kilometer against the stream in $11\frac{1}{4}$ minutes and down the stream in $7\frac{1}{2}$ minutes. The speed (in km/hr) of the man in still water is: (Answer – 5)

79. A man takes twice as long to row a distance against the stream as to row the same distance in favour of the stream. The ratio of the speed of the boat (in still water) and the stream is : (Answer – 3:1)

80. A man rows to a place 48 km distant and come back in 14 hours. He finds that he can row 4 km with the stream in the same time as 3 km against the stream. The rate of the stream is: (1Km/hr)

Index

22786218R00115

Printed in Great Britain
by Amazon